# DARWIN'S ARMADA

ALSO BY IAIN McCALMAN

*Radical Underworld: Prophets, Revolutionaries and Pornographers in London, 1795–1840*

*Horrors of Slavery: The Life and Writings of Robert Wedderburn*

*The Seven Ordeals of Count Cagliostro: The Greatest Enchanter of the Eighteenth Century*

IAIN McCALMAN

# DARWIN'S ARMADA

*Four Voyages and the Battle for the Theory of Evolution*

W. W. NORTON & COMPANY
*New York • London*

First American Edition 2009



Manufacturing by Courier Westford
Production manager: Anna Oler

Library of Congress Cataloging-in-Publication Data

McCalman, Iain.
Darwin's armada : four voyages and and the battle for the theory of evolution / Iain McCalman. — 1st American ed.
p. cm.
Includes bibliographical references and index.
ISBN 978-0-393-06814-6 (hardcover)
1. Evolution (Biology)—History—19th century. 2. Darwin, Charles, 1809–1882. 3. Hooker, Joseph Dalton, Sir, 1817–1911. 4. Huxley, Thomas Henry, 1825–1895. 5. Wallace, Alfred Russel, 1823–1913. I. Title.
QH361.M37 2009
576.8'209034—dc22 2009016055

W. W. Norton & Company, Inc.
500 Fifth Avenue, New York, N.Y. 10110
www.wwnorton.com

W. W. Norton & Company Ltd.
Castle House, 76/76 Wells Street, London W1T 3QT

1 2 3 4 5 6 7 8 9 0

*To Kate and Rohan*

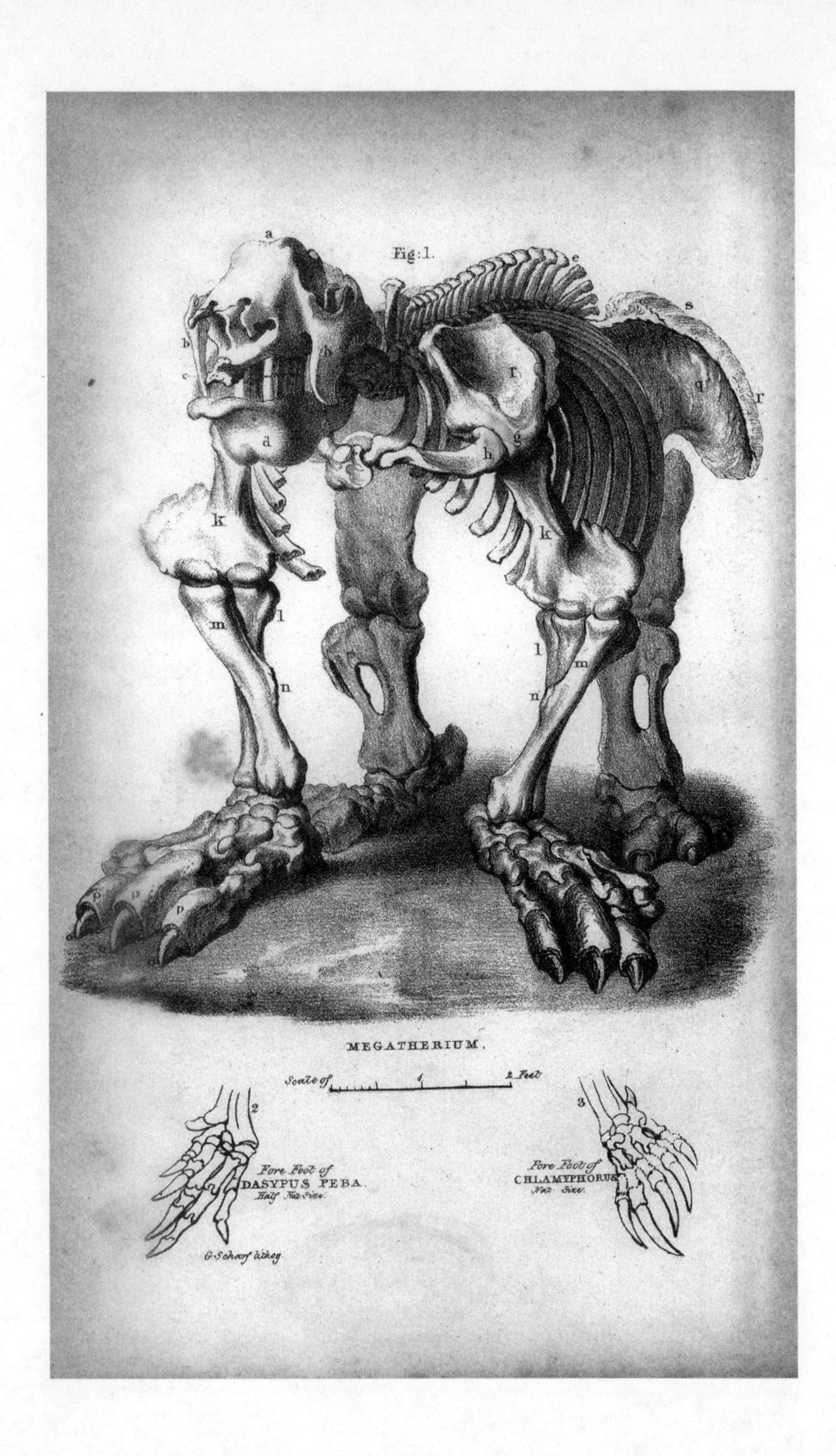


MEGATHERIUM.

# Contents

PART FOUR

## Alfred Wallace in the Amazon and South-East Asia, 1848–66



PART FIVE

## The Armada at War, 1859–82

# Prologue:
# Darwin's Last Voyage

*Oh build your ship of death, oh build it! for you will need it.*
*For the voyage of oblivion awaits you.*
D. H. Lawrence, 'The Ship of Death'

Charles Darwin's funeral took place at Westminster Abbey on Wednesday 26 April 1882. Twenty years earlier, the English press had taunted him as 'The Devil's Disciple', the scientist whose theory of evolution had dethroned the divine creator and turned man into the cousin of the monkey. Now the *Pall Mall Gazette* spoke for all in comparing him to Copernicus and calling him 'the greatest Englishman since Newton'. The more than two thousand mourners at the Abbey made up a Who's Who of the Victorian establishment. So many had applied for admission cards that the undertakers were rattled.

The body had arrived at eight o'clock the evening before, after a horse-drawn journey from the village of Downe, in Kent, accompanied for the sixteen miles by three of Darwin's sons and an icy drizzle. The white oak coffin, bearing the simple inscription *Charles Robert Darwin, Born February 12, 1809. Died April 19, 1882*, was carried into the dim lamplight of the Abbey's Chapel of St Faith, where it perched like a small ship in dry dock.

Shortly before eleven o'clock the next day, the Darwin family, friends and a few servants made their way into the Jerusalem Chamber. Dignitaries took their positions in the Chapter House, and the choir assembled in the stalls. The pews on the south side of the nave registered the whispers of frock-coated scientists, philosophers, admirals, ambassadors, museum directors, politicians, philanthropists, civic worthies, university professors and clergymen. Now the non-ticketed seats in the north-western part of the nave and the back rows began to fill with ordinary folk, some plain curious, some keen to pay homage to the man who'd once shaken the foundations of the Church. They included a sprinkling of old radicals – Chartists, republicans, and freethinkers like G. J. Holyoake – for whom Darwin's ideas had been an inspiration.

A few pointed absences among the country's mighty were noted, though each claimed an excuse: Queen Victoria was busy preparing for her son Prince Leopold's wedding; Prime Minister Gladstone, a fervent evangelical with no love of Darwin's ideas, was caught up in the political mire of the Irish independence struggles; the Archbishop of Canterbury was 'indisposed'; and the Dean of Westminster Abbey was 'abroad'.

Other representatives of the Church of England made up for the timidity of their clerical seniors: canons, vergers and clerks were present in abundance. As the bell tolled noon, the Queen's Chaplain-in-Ordinary, George Prothero, opened the ceremony with the song 'I Am the Resurrection', glossing over Darwin's well-known scepticism about life after death. Everyone knew that he had rejected the idea of a divine creator who'd intelligently designed the world of man and nature: he believed only in nature's implacable laws of change, chance, struggle, survival and extinction.

The coffin, a black velvet coverlet draped over it, carried two wreaths – lifebuoys for another world – and a spray of white blossom at the prow. Ten pallbearers bobbed it slowly up the nave to its resting place at the northern end of the choir screen, close to the statue of Sir Isaac Newton. Three of the pallbearers – two dukes and an earl – represented the state and Cambridge University, where Darwin had once been a clerical student. Ambassador James Russell Lowell, another bearer, represented the United States of America. Sir William Spottiswoode, President of the Royal Society, was there for the scientific establishment. Darwin's neighbour and friend, Sir John Lubbock, a Liberal MP, London banker and distinguished amateur archaeologist, embodied Victorian government, finance and culture.

Three men in late middle age who also gripped the brass handles of the coffin were there because they'd been Darwin's closest friends and intellectual collaborators. Biologist Thomas Henry Huxley, aged fifty-seven, was tall and thickset with a beaklike nose and massive side whiskers. Next to him stooped botanist Joseph Dalton Hooker, sixty-five, slight and fragile with a leonine ruff of white hair circling his face, which was pale from angina. At the rear was zoogeographer Alfred Russel Wallace, fifty-nine, tall and gangly with a heavy white beard around a kindly mouth.

The modest co-discoverer of the theory of natural selection, Wallace was a man whom Darwin had revered, despite Wallace's reputation for radical eccentricity. Privately, George Darwin thought it would have been more in keeping with his father's feelings to have positioned Wallace 'at the other end' of the coffin with his two colleagues.[1] These three scientists belonged together: they were not only the dead man's most committed scientific supporters, but, as fellow southern voyagers, they had also shared with him a special bond of the 'salt'.

As if to remind them of that formative period in all their lives, Darwin's coffin came to rest next to another southern traveller, whom Darwin had first met at the Cape of Good Hope half a century earlier. This neighbour in death was the eminent astronomer-philosopher

Sir John Herschel.[2] The two old travellers nudged each other in the confined space of the Abbey. As a seaman, Darwin had been familiar with constrictions of space. One of his mourners, Admiral John Lort Stokes, had written to *The Times* five days earlier to tell readers how he had worked beside Darwin on the *Beagle*'s poop-cabin table while their hammocks swayed overhead. Sudden bouts of seasickness would force Darwin to leave his microscope and lie down, saying, 'Old fellow, I must take the horizontal for it.'[3] And here he was taking the horizontal for the last time.

The organ sounded a final anthem, Canon Prothero pronounced the Benediction, and Charles Darwin, that most reluctant sailor and fighter, embarked on his last voyage to meet the worms he'd been so recently studying.

Locals from Darwin's tiny Kentish hamlet of Downe were represented by two long-time family servants, Mary Evans, who'd looked after Darwin since he was a boy, and old Mr Parslow, the almost equally long-serving butler. Provision had been made for other villagers to attend the funeral, but none did so. Most believed that Darwin would have hated the Abbey ceremony. Had it been held at the local church of St Mary's as they'd hoped, and as he'd intended, the pallbearers would not have included all these stiff society folk. Family members, neighbours, and old friends like Hooker, Huxley and Wallace would have carried the coffin to the resting spot Darwin had requested.

The Downe church was small, just a nave and a chancel below a white plaster ceiling and an old timbered roof, but it was dignified in its simplicity. The dwarf tower and tall spire were the first thing travellers glimpsed through the horse chestnut trees as they made their way down from Keston Hill, along a steep winding path cut into the chalk, then across a level meadow and into the village. A huge old elm with a 23-foot girth shadowed the entrance to the church and the adjacent yard.[4] This was the burial spot of Darwin's two children who died in infancy, Mary Eleanor and Charles Waring. Here too, in a grassy patch under a large old yew tree, lay two of Darwin's Wedgwood cousins, sharing with his elder brother Erasmus a specially commissioned family vault designed to hold twelve. In spite of Charles's lack of religious feeling, he'd felt this tiny churchyard to be 'the sweetest place on earth'. So when his stuttering heart finally stopped beating at four pm on 19 April, after a week of pain, both the family and the village took it for granted that a quiet local ceremony would be held. Darwin had expressed this wish to his wife Emma a year earlier, when gripped by an intimation of death.

John Lewis, the Downe village carpenter and joiner, used his father's tools to make Darwin's coffin.[5] Years earlier, Lewis's father had helped renovate Charles and Emma's shabby country house. John Lewis had established an oddly intimate relationship with the famous naturalist at the age of fourteen, when he was hired to slosh buckets of ice-cold water over Darwin's shivering frame as part of a modish water cure for the bowel problems that had made him a semi-invalid. Like so many of the villagers, Lewis respected Darwin as an eccentric but generous supporter of the local poor. They shook their heads when telling stories of seeing him stare for hours at an ant's nest. He would gravely thank the local children who rushed to open the gate for him as he took his afternoon walk. Lewis had also made the coops for Darwin's beloved pigeons, which seemed to occupy more of his time even than ants and barnacles. And it was probably the young carpenter who also built the special 'invalid chair' that Darwin was compelled to use in his last months to get up the stairs to his bedroom.[6]

The coffin was built with strong planks of local oak. 'I made his coffin just the way he would have wanted it,' John Lewis later recalled, 'all rough, just as it left the bench, no polish, no nothin.'[7] He loaded it onto his cart and trundled up to the house, where he helped the family to lay Darwin out in the local way. The dead man looked lifelike and serene, a sight so poignant that it set his three grown-up sons, William, George and Francis, sobbing. Their grief in turn breached their mother's crust of control. Emma, at seventy-four, had known her first cousin Charles Darwin all her life, and she had spent forty years of their marriage nursing his ailments, worrying about his godlessness and protecting his work habits. Tough as she was, she sat down beside his coffin and cried.[8]

As Darwin's great biographer Janet Browne observes, dying is often our most political act. It was so with Darwin, and none knew this better than Thomas Huxley, the most political and combative of Darwin's disciples. Huxley thought it appropriate to the historical stature of his friend, as well as essential for the cause of professional science, that he be buried at Westminster Abbey. It was the memorial site of Britain's greatest achievers: '. . . 50 or a 100 years hence it would seem absolutely incredible to people,' Huxley argued, 'that the state had in no way recognized his transcendent services to Science'. Under his coaching, the *Standard* newspaper went further still; it implied that the Darwin family had a patriotic duty to accede to the people's wishes. 'We owe it to posterity to place his remains in Westminster Abbey, among the illustrious dead,' the editor thundered.

This was only one of a barrage of tactical moves that Huxley devised in order to engineer an appropriately grand funeral. Joseph Hooker, his usual helpmate in such matters, was unfortunately 'utterly unhinged' by Darwin's death.[9] And despite his own magisterial scientific stature, Huxley faced some formidable obstacles, not least being to persuade some of Britain's most senior Church of England clergy to bury a professed religious sceptic in the Abbey. Whatever their own private ideas on evolution – and many clergy had found ways to reconcile it with a divine creator – they could not appear to be encouraging freethinkers and atheists. But Britain's most eminent scientific writer and educator had not acquired his newspaper nickname of 'Pope Huxley' for nothing. He was as wily as a pontiff. He induced a clergyman friend, Reverend Frederick Farrar, the Canon of Westminster and an amateur naturalist of liberal views, to put the case for a Westminster burial to the Dean of the Abbey.

Then there was the reluctant family, who were anxious to carry out Darwin's wishes and did not want to be thought vainglorious. Sir John Lubbock was recruited to persuade them. He wrote to William Darwin admitting that privately he too would have preferred to see Darwin buried 'in our quiet little churchyard where someday I would have joined him', but he emphasised that 'from a national point of view it is clearly right that your father should be buried in the Abbey'.[10] Emma was adamant that Darwin's funeral not be spoilt by the religious controversy that had dominated the press following the publication of *The Origin of Species* in 1859. Sir William Spottiswoode assured her there'd be none, and once her eldest son decided in favour of Huxley's proposal, Emma acceded readily enough.

Politicians, both Liberals and Tories, had also to be coaxed, especially Liberal Prime Minister Gladstone, who had frequently clashed with Huxley over evolution. Here too Huxley sought the help of Lubbock, who lobbied a group of science-friendly politicians of all parties and obtained a petition in support of a Westminster Abbey funeral. Public opinion had to be stimulated and guided, but not too obviously. Huxley chose to work through the *Standard* because of its impeccably conservative reputation.

Finally, some of Darwin's important fellow scientists and friends had to be rallied. Joseph Hooker, for example, hated any type of ceremonial fuss and favoured a quiet local event.[11] Alfred Wallace lived so quietly in the countryside these days that Huxley forgot about him altogether, until reminded by one of Darwin's sons. Cursing himself for his stupidity, Huxley hastily invited him to be a pallbearer.

To Huxley's relief, all the leading newspapers followed the cue of the *Standard* in arguing that British patriotism demanded a suitable acknowledgement by the state of Darwin's achievements, especially when countries like Germany and France had long accorded him their highest honours. The Vienna *Allgemeine* had been typical in declaring: 'Our century is Darwin's century, we can suffer no greater loss'.[12] This cosmopolitan ownership of Darwin had to be usurped; Britain's national reputation was at stake. Even newspapers like *Clerical World* agreed that the days of the evolution wars were long forgotten; an accommodation between Christianity and evolution was possible. So successful was Huxley's strategy that *The Times* thought it was telling the truth when it claimed that the idea of an Abbey funeral 'arose, not apparently, in any single mind, but spontaneously and everywhere . . . it was felt that the Abbey needed it more than it needed the Abbey. The Abbey tombs are a compendium of English deeds and intellect. The line would have been incomplete without the epoch-making name of DARWIN'.[13]

With clerical, political, public and family agreement finally secured, Huxley set about organising the trappings for a grand Abbey funeral. A famous firm of Piccadilly undertakers, Messrs T. and W. Banting of St James Street, was hired; they'd been the orchestrators of one of the century's most magnificent state funerals, that of the military hero Wellington in 1852. John Lewis's rough coffin was discarded. 'They sent it back,' the carpenter complained, and shunted Darwin into a new one, 'so shiny you could see to shave in.'

Lewis and some other villagers, including the aggrieved publican of the George and Dragon, did share an element of self-interest in their grumbling. Having the world-famous naturalist buried in Downe would be excellent for business: the village would become a place of pilgrimage, Lewis's coffins would be legendary and pints of beer in brisk demand. But the locals had a point when they complained about the cavalier way in which Darwin's wishes had been overridden. Lewis was right in saying that the reclusive Charles Darwin 'always wanted to lie here, and I don't think he'd have liked [a Westminster Abbey burial]'.[14]

There was a deeper sense, too, in which Darwin had a right to be recognised as a Kentish naturalist. It was not just that he'd lived for forty years in that brick house with its azaleas and giant old mulberry tree; that he'd conducted his famous homely experiments using vines, floating seeds, and pigeons in its hothouses and roosts; that he'd written most of his great books in its cluttered, comfortable study with engravings of Hooker and Huxley on the wall; or that he'd exchanged ideas in the garden with some of the most brilliant scientists in the world. All these were important enough reasons for him to be buried in the village, but some newspapers also hinted that Darwin had become so intimate with the 'natural economy', as they termed it, that it seemed wrong to remove him from it.

Darwin had been linked to the bucolic father of naturalism, Reverend Gilbert White, whose minute and loving journal of the daily activities and interdependencies of the plants and animals of the village of Selborne was a classic.[15] Darwin had made a pilgrimage with his son Francis to White's house and garden in Surrey, and had begun but never completed 'An Account of Downe', which Francis thought was intended to be a natural-history diary in the style of White's.

Like Gilbert White, Darwin had walked every inch of the fields, meadows and birch woods surrounding his home, often several times a day, observing insects, birds and plants with an eye made keen by years of trained observation. Though nobody understood better than he the savagery of nature, nobody better appreciated the 'peace and silence' of the hamlet that visitors so often remarked upon.[16] He'd recognised that Down House, though only fifteen miles to the south-east of London, was not unlike a remote island in the Pacific, 'absolutely on the verge of the world'.[17] This made it an ideal natural laboratory for testing his ceaseless flow of ideas. His final book, on earthworms, had not only been inspired by a passage in White's *Natural History of Selborne*, but had also resulted from his work on the soil and moulds in his own garden.[18]

In the end, despite agreeing to the Abbey funeral, Emma decided not to attend. Privately she admitted that 'it gave us all a pang not to have him rest quietly by [his brother] Eras', and she preferred to stay at the house, which was cluttered with Darwin's objects and suffused by his presence. A visiting journalist had felt the same sensation as he looked at the shawls, cloaks and cushions still sprawled on Darwin's 'ungainly' chair in front of the fire, at the books, papers, glass shades and card boxes scattered on his table, and at the two plants on a low bench that he'd been studying just before he died.[19]

The boys and their wives went to Westminster Abbey, where William showed that he'd inherited not only his father's propensity to baldness, but also his hypochondria. Feeling a breeze blowing through the Abbey and fearing a chill, William sat throughout the ceremony with his gloves perched on his head. Reverend Farrar later told William that the magnitude of the public tribute had 'overshadowed private feeling'.[20] Neither did the family entirely escape the controversy that Emma had feared. Lady Hope, an aptly named evangelical who'd preached at Downe towards the end of Darwin's life, began a rumour that he'd rejected evolution on his deathbed and declared himself a Christian. It was nonsense of course – - Darwin's last words had actually been 'I am not in the least afraid to die' – but the resultant newspaper fuss distressed Emma and the children.

Deep down, those three scientists who had helped carry Darwin's coffin probably agreed with his wife; each would have preferred a more intimate local burial for the man they had worked with and loved. But they also saw that his funeral was the last and perhaps greatest symbolic battle in a war they'd been fighting for twenty years. They had waged a fierce, unremitting campaign in defence of Darwin's theory of evolution by natural selection, against a host of scientists, philosophers, clergymen, politicians and others who found its irreligious implications unacceptable. They had fought, too, for the kind of observational, fact-gathering and law-making mode of science that Darwin pioneered.

Thomas Huxley, biologist, palaeontologist and social philosopher, had been tireless in taking Darwin's ideas to the public, and in extending the implications of evolution to other spheres of science and culture. In private he mourned Darwin as intensely as anyone: he felt 'a poignant grief . . . to think I shall never hear his cheery welcome or feel his cordial grasp again'.[21] Speaking in public, though, Huxley still used the language of war. He extolled Darwin as a hero and a fighter who 'led us to victory . . . He found a great truth, trodden underfoot, reviled by bigots, and ridiculed by all the world; he lived long enough to see it, chiefly by his own efforts, irrefragably established in science, inseparably incorporated within the common thoughts of men . . .'[22]

Alfred Wallace, though in some ways as shy and retiring as Darwin himself, echoed Huxley's martial sentiments. He was proud to be called Darwin's 'true knight', the man who had independently arrived at the same theory of natural selection before Darwin made his own public, and who chivalrously accorded all priority and fame to the older man. Wallace thought of himself as merely a 'guerrilla chief' in the evolution war, 'while Darwin is the great general, who can manoeuvre the largest army, and . . . lead on his formation to victory'.[23]

Joseph Hooker, botanist and biogeographer, had been an equally fierce and effective warrior for natural selection, though he took greatest pride in being the man whose judgement Darwin trusted absolutely: 'my public and my judge'. As he made his last farewell over Darwin's open grave, Hooker felt saddened to be one of only a few men who really knew the reclusive scientist intimately. 'But for those of us who have now to mourn so unspeakable a loss, it is some consolation to think, while much that was sweetest and noblest in our lives has ended in that death, his great life and finished work is still before our view; and in regarding them we may almost bring our hearts to cry – Not for him, but for ourselves we weep.'[24]

And weep they did – for their present, their future and their past. It is so often forgotten that what had brought these four very different and distinguished Victorian figures together so as to be 'strengthened in brotherly love' was their separate participation as young men in daring scientific voyages of exploration to the southern oceans.[25] These four voyages created 'a Masonic bond' as a result of being 'well salted in early life'.[26] The voyagers were tested, emotionally, physically and intellectually, and they felt themselves transformed in the deepest sense – as scientists and as people.

The three younger men had each been conscious of following in Darwin's wake. His famous *Voyage of the Beagle*, detailing his trip to South America, Australia and the South Seas, had been their inspiration. It offered them so much, beginning with what Darwin modestly called 'a general interest in Southern lands': poetic descriptions of tropical landscapes, exciting stories of adventure, exhilarating new methods of discovering the forces that shaped the physical and biological habitats of the Southern Hemisphere. For as the *St James Gazette* rightly observed, 'Mr Darwin was not only a discoverer, but a captain and organizer of discovery.'[27]

Through their South Seas odysseys, these four young, romantically minded amateur naturalists gained access to one of the richest natural laboratories on the globe. They each discovered evidence from which to build new scientific theories, and each stored lifelong memories of a common experience of hardship and pleasure that bound them together like shipmates. Out of these southern adventures grew their friendship, their interlocking scientific interests, and finally, their joint participation in Darwin's evolution war. The southern oceans were the training ground of the seamen who would lead Darwin's armada to ultimate victory.

THE GRAPHIC

AN ILLUSTRATED WEEKLY NEWSPAPER

No. 649—Vol. XXV. Regd. at General Post Office as a Newspaper

SATURDAY, MAY 6, 1882

WITH EXTRA SUPPLEMENT

PRICE SIXPENCE Or by Post Sixpence Halfpenny

BAZAAR AND FANCY FAIR IN THE EXHIBITION PALACE, DUBLIN, IN AID OF THE MASONIC FEMALE ORPHAN SCHOOL

FUNERAL OF THE LATE CHARLES ROBERT DARWIN IN WESTMINSTER ABBEY

# PART ONE

# Charles Darwin and the *Beagle*, 1831–36

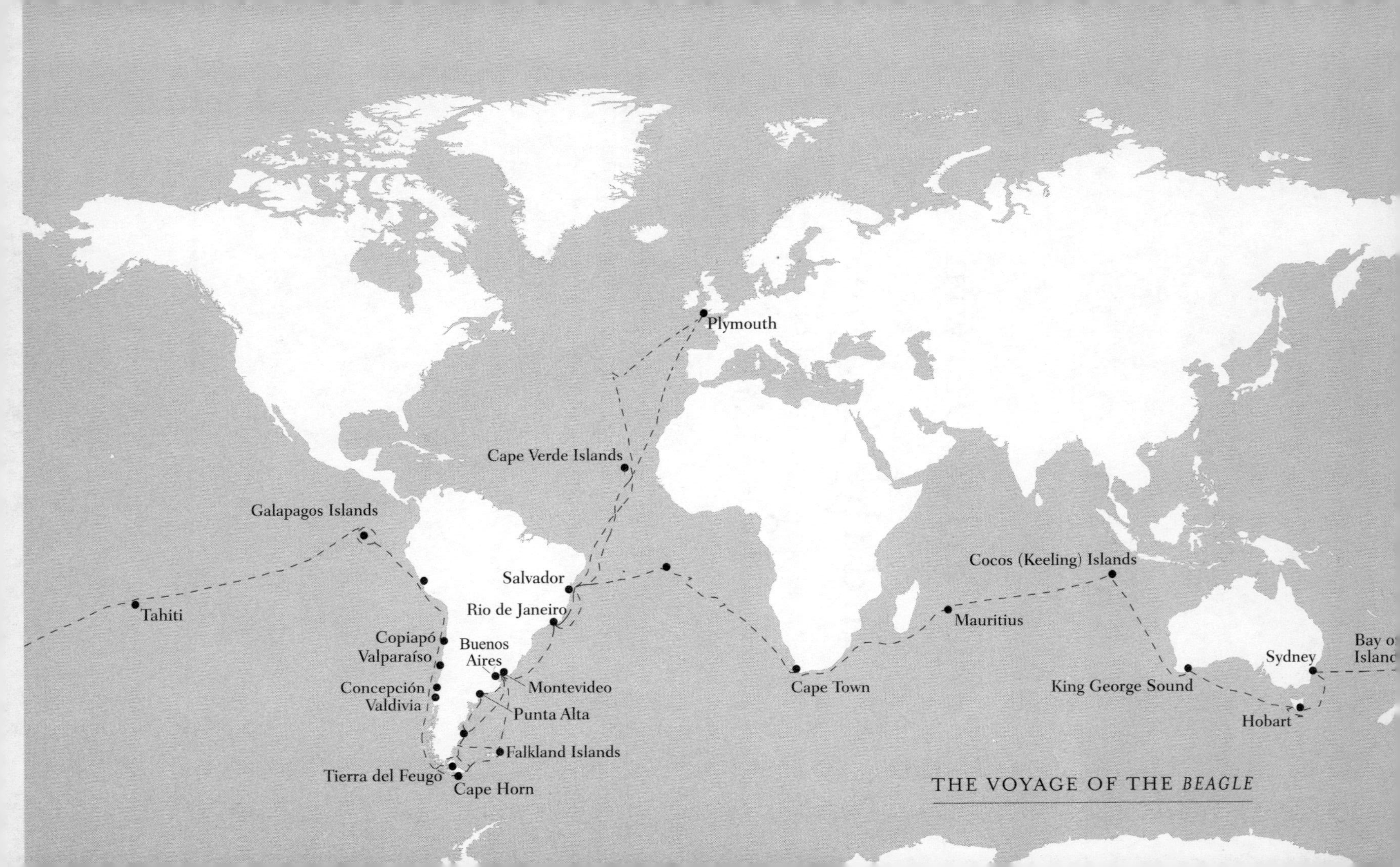
Plymouth
Cape Verde Islands
Galapagos Islands
Salvador
Tahiti
Rio de Janeiro
Copiapó
Buenos
Aires
Valparaíso
Concepción
Valdivia
Montevideo
Punta Alta
Falkland Islands
Tierra del Feugo
Cape Horn
Cape Town
Mauritius
Cocos (Keeling) Islands
King George Sound
Sydney
Hobart
THE VOYAGE OF THE *BEAGLE*

# The Prodigal Son

On 29 August 1831, Charles Darwin, aged twenty-two and just returned home from Cambridge University for two weeks of holiday, presented his father with a ridiculous proposition. Charles had a letter from a Cambridge friend, Reverend John Henslow, conveying an offer of a position as naturalist on a naval survey vessel that would depart England in one month's time, to circumnavigate the globe via Tierra del Fuego and the East Indies. The voyage would take two years and Charles would have to pay for himself.

It wasn't the cost that Dr Robert Waring Darwin objected to: heaven knows, he was used to covering his son's extravagances. It was rather an exasperated fear that the boy would end up 'an idle, sporting man'. He worried that Charles had been spoiled. Perhaps his dreamy, aimless character was a result of the early death of his mother, in 1817 when he was only eight, and his upbringing by three sisters who'd indulged and scolded him in equal measure. Charles had shown a childhood tendency to draw attention to himself by making up romantic and sensational stories. He'd also developed a passionate love of the outdoors.

As a doctor and a man of robust commonsense, Robert Darwin knew that a boyish enthusiasm for fishing, raiding birds' nests, stealing fruit, collecting shells and hunting rats was perfectly natural and healthy. It had shown to good effect in Charles's tall, muscular frame and strong legs. Dr Darwin, who kept a detailed record book of his botanical plantings, had also gained great pleasure from his younger son's company in the garden of their Shrewsbury house, the Mount. On the other hand, the doctor had begun to realise, unless Charles's recreational zest was balanced by a willingness to commit to some form of useful learning, it could become a means of escaping the responsibilities of adulthood.

Dr Darwin wondered if he was partly to blame for Charles's lack of direction. Perhaps he'd been too preoccupied with work and too generous with the boy's allowance? The doctor had been so devastated by the loss of his wife, Susannah, that he'd immersed himself in his profession. Being the best-known, best-paid physician in Shrewsbury was demanding of his time, and he'd also developed a supplementary business as a money broker: he raised money for investors and occasionally invested personally in the burgeoning network of canals and roads that were criss-crossing the industrialising county of Shropshire. This could be risky, of course, but it was also highly profitable, as long as one was careful and a good judge of character. Dr Darwin was both, and he'd become one of the richest men in the county.

Charles was a puzzle. He'd grown up a delightful boy in so many ways – affectionate, sensitive and energetic. He listened attentively to his elders, doted on his sisters, Marianne, Caroline, Susan and Emily, and adored his clever, rather literary older brother Erasmus. Neither did Charles seem stupid; if he put his mind to something he was able to learn quickly enough. This hadn't shown in his school results, admittedly, which were average to mediocre. At the age of nine he'd been sent as a boarder to Shrewsbury Grammar, a school close to their house and famous throughout the country under the headmastership of the learned Dr Samuel Butler. Although Charles had appeared to work conscientiously enough, he showed no talent for the classical studies that dominated the school's curriculum. He'd been chastised by the headmaster for his lacklustre efforts in Latin and Greek verse, and for wasting time reading Shakespeare's histories and fooling about with chemicals. His nickname of 'Gas' was not intended to be complimentary.

Charles had never settled at the school; he'd taken advantage of every unmonitored period in the daily routine to dash the mile home to gossip with his sisters and play with his dogs. When not making these risky little escapes, he mooned about reading poems and romances by Byron and Scott, or sensational travel and adventure stories in boys' magazines. He had a vivid imagination and something of an aesthetic bent, which showed up in his stories, his reading and his 'poetic fancy'.[1] He also developed a passion for hunting on his uncle Josiah Wedgwood's country estate, Maer, thirty miles from Shrewsbury.

By the time Charles was fifteen, Dr Darwin decided there was no point in him continuing at the school. Tolerant as he normally was, the doctor made his first of several angry predictions: 'you care for nothing but shooting, dogs and rat-catching, and you will be a disgrace to yourself and all your family'.[2] He sent Charles to Edinburgh University to become a physician, in the footsteps of his grandfather, father and older brother. Although the Darwins boasted a long and distinguished lineage as country gentry, they prided themselves on acquiring respectable vocations. Erasmus, who was finishing his medical degree at Edinburgh, would help Charles find his feet.

Robert Darwin had already seen some hopeful signs that his younger son might apply himself in this capacity. As a child, Charles loved to squeeze himself into the carriage beside his father's massive bulk and listen to his medical theories and diagnoses. In the months immediately before going to university, Charles showed that he'd grasped Dr Darwin's central tenet of good doctoring: be optimistic and confident and a cure will often follow; most patients were not really seeking medical treatment so much as advice and comfort about their domestic unhappinesses.[3] Charles tried this remedy on a small group of patients among the local poor, who'd been pleased by the attention. Dr Darwin was impressed by his son's sensitive manners and careful records.[4]

But the Edinburgh experiment proved a disaster, though it took nearly two years before Dr Darwin realised the full extent of his son's alienation. Despite enjoying the convivial company of Erasmus and his friends for the first year, Charles complained in letters to his sisters about musty, boring lectures on the *materia medica*, and disgusting, blood-soaked anatomy demonstrations. Here again Dr Darwin was not unsympathetic: he hadn't liked medicine at first either. He understood his son's repulsion at having to dissect corpses amidst the stench of guts, and his distress at having to watch limbs being hacked off screaming children, but Dr Darwin had learnt to tamp down his squeamishness and do the necessary minimum of these activities.

Charles, by contrast, let his feelings run away with him and swore never to lay eyes on such horrors again. Dr Darwin had given both his sons to understand that if they exerted themselves in order to gain a medical qualification, they wouldn't necessarily have to practise: they would inherit enough money to live in comfort. Erasmus had taken the hint, but Charles seemed to treat the prospect of an inheritance as a further inducement to idleness.

There was little consolation for the doctor in Charles's extracurricular enthusiasm for naturalism. Natural history and philosophy were fine as adjunct subjects to a medical degree, but they would not in themselves lead to any paid employment. Naturalism was mostly the preserve of enthusiastic amateurs – clergymen whiling away idle moments in their rural parishes, genteel young women drawing butterflies and pressing plants, artisans finding themselves a self-improving hobby. Lectures in subjects like botany were given at Edinburgh mainly to young men intending to practise in the colonies, where medicines were not easy to obtain.

And it was not as if Charles was any more conscientious about attending lectures in those fields than in other parts of the curriculum. He griped to his sisters about the tedious zoological and geological lectures of the famous scholar, Professor Jameson, and he swore to never again study or read a book on the wearisome subject of geology.[5] No, Dr Darwin suspected that his son's fascination for natural history was similar to his obsession with hunting – commendable for its zeal and harmless as a hobby, but an undesirable substitute for serious work. Though generous about covering a stream of expenses, the doctor found it difficult to see how some of them – such as paying a former black slave to teach Charles how to stuff birds – were going to advance the boy's career.

Charles was also prone to wandering along Scottish beaches with naturalist-minded professors, gathering up crustacea in tide pools, and philosophising about the age of the earth. Dr Darwin – whose own father, another Erasmus, had been a famous poet as well as a physician – understood the pleasures of the imagination, but he didn't fool himself that going trawling for oysters with Newhaven fishermen qualified as work. Such activities looked much more like an extension of Charles's delight in stalking partridge through the woods, or tramping across the north Welsh countryside to admire crashing waterfalls.[6]

When Charles eventually gave up his medical course in 1827, Dr Darwin felt bound to put his foot down. If the boy wanted to fritter away his life shooting snipe and collecting beetles, he must at least take a degree that would cloak such hobbies with a modicum of gentility. In short, he must resort to the stock refuge of the educated idle and become a Church of England clergyman. True, the Darwins were not a religious family – Charles's mother Susannah Wedgwood had been raised a Unitarian, a nonconformist religion so rationalist as to border on free thought – but, largely for social reasons, her children had been baptised in the Church of England. Charles's easygoing orthodoxy extended to believing vaguely in the truth of the Bible and adhering to the Anglican creed. So it was agreed that he would go to Cambridge to take a bachelor's degree.

Everyone in the family knew that his incentive was not so much religion as an aspiration to emulate the life of the famous clergyman naturalist, Reverend Gilbert White, author of *The Natural History of Selborne*, a gentle, arcadian record of the daily doings of the plants, birds and animals of a small rural parish in Hampshire. After reading it, Charles had wondered excitedly 'why every gentleman did not become an ornithologist'.[7]

Charles's reduced career goal proved within range of his stamina and abilities. He struggled at first: the wasted years at school meant that he had to take special tutoring to regain some rudimentary knowledge of Latin and Greek and to brush up his algebra. Once at Cambridge, he continued to party hard, but he also worked enough to pass his preliminary exams with ease, and eventually to graduate in the respectable position of tenth out of the non-honours candidates.

Dr Darwin's relief was so palpable he was happy to indulge his son's leisure activities. He didn't jib greatly at Charles's extravagant spending on food, cards and wine, or at reports from his daughters of the boy's continued obsession with hunting, rambling and insect collecting. On one occasion, his tutor at Christ's College was greatly alarmed by cracking sounds emanating from behind young Darwin's door: Charles was practising snuffing out a moving candle with the puff of air detonated by his unloaded shotgun.[8]

At Cambridge, Charles met up with a former teacher of his brother's, the Reverend John Henslow, who inspired him to become a regular at Friday soirées and weekend naturalist outings – and even to attend a course of botany lectures. Through Henslow, Charles mixed with senior university dons like William Whewell, the natural philosopher, and Adam Sedgwick, the geologist.[9] Charles's hero-worship of the sociable and tolerant Henslow appeared to be all to the good. Aside from provoking him into becoming an even more energetic gatherer of plants and insects, it reinforced his belief in the congeniality of a clerical career.[10]

Dr Darwin did not even mind when Charles, towards the end of his time at Cambridge in 1831, read a book lent to him by Henslow and became infatuated with the idea of exploring the tropical environment of Tenerife in the Spanish-owned Canary Islands. It was an eccentric destination, but the boy deserved a holiday as a reward for attaining his degree.

In the event, Charles's zeal outran his planning abilities, and the trip was delayed for a year because of a missed schedule and the death of one of his intended companions. Henslow, in the meantime, gave constructive direction to Charles's naïve enthusiasm by urging him to prepare for the trip intellectually. Though hopeless at languages, Charles laboured for a while to teach himself the rudiments of Spanish, and thanks again to Henslow's good offices, undertook a short field trip in August 1831 as assistant to Adam Sedgwick. The gruff Yorkshire professor agreed to give the boy a crash course in practical geology among the mountains of Snowdonia in North Wales, so that he might better understand the volcanic structures of Tenerife.

It was immediately after this episode that Henslow's note about the voyage around the world arrived. Dr Darwin pointed out in no uncertain terms that this was a wholly different prospect to holidaying in Tenerife for a few weeks: it could involve years of absence, along with the perils of shipwreck, drowning and tropical disease. The doctor knew no more about sailing than did his son, but he did know that sailors were regarded as social vagabonds who spent all their time on shore drinking, whoring and fighting. Keeping such company could only harm Charles's reputation and his chances of gaining a clergyman's position, along with his commitment to such a career. As for working as a ship's naturalist in South America, what possible good could it do, especially when Charles had precious few credentials for the job?

It was obvious, Dr Darwin speculated, that other, qualified naturalists had already turned down the offer. Had Charles thought to ask why? What was it about the captain, the ship or the proposed voyage that had led them to refuse? And what would it say for Charles's future standing that he was willing to take on a position spurned by better men?[11]

Dr Darwin's instinct was correct: there was something unusual about the captain, the voyage and the ship, still more about the request to take on his son as naturalist. He was wrong, however, to assume the worst. Captain Robert FitzRoy's proposed voyage of the *Beagle* to survey South America and other parts of the Southern Hemisphere was distinctive mainly for being one of the best-backed, best-prepared and most ambitious survey voyages sponsored by the Admiralty Board since the end of the Napoleonic Wars in 1815.

Captain FitzRoy, though only twenty-six years old, was one of the stars of Britain's peacetime navy. In an institution where social rank still mattered as much as ability, his credentials were stellar. Descended from the ennobled bastard line of Charles II, he counted the Duke of Grafton as his uncle on one side of his family, and Lord Londonderry, brother of the late Viscount Castlereagh, on the other. Both uncles were influential in court and government. Because his father, Lord

Charles FitzRoy, owned large swathes of land in Northamptonshire, Robert was also personally wealthy. On top of all this, he'd benefited from the modernising impulses within the British navy. After being educated at Harrow, he graduated through the Royal Naval College at Portsmouth with record marks, excelling in subjects like mathematics, astronomy, mechanics, hydrostatics and cartography.

Once at sea, he proved an equally brilliant practical sailor. He honed his sailing and navigation skills as a junior officer on stints in the Mediterranean, in South American waters, and along the treacherous Cornish coast, before eventually catching the eye of an ex-Nelson veteran, Sir Robert Otway. The flag-lieutenant's influence soon procured FitzRoy's first command, in October 1828 – yet it had come under the most testing of circumstances.

A few months earlier, Captain Philip Parker King's naval surveying expedition in South America had been forced to make a sudden return to Rio de Janeiro because the captain of his support vessel, the *Beagle*, had shot himself in a fit of depression. Captain Pringle Stokes had recorded in his log plenty of reasons to be depressed. He had been charged with mapping the coastline of bleak Tierra del Fuego, near Cape Horn, and his ship was dangerously unseaworthy, his charts riddled with errors, and his men ill with scurvy. Tierra del Fuego had the same latitude as Newfoundland but worse weather. Incessant grey sleet and violent storms battered the *Beagle* and frayed the captain's nerves. Under such conditions, he confided to his journal, 'the soul of a man dies in him'. He decided his body should follow suit.[12]

Following the collapse of the Spanish and Portuguese empires during the Napoleonic wars, a rash of independent nations had emerged in South America. While the United States proclaimed their authority over the area in the Munro Doctrine, European sea powers jockeyed to exploit the opportunities opened up within this resource-rich region. The British navy surveyed much of the continent's east coast, paving the way for trade by enhancing the navigability of one of the most dangerous coastlines in the world and justifying the presence of substantial peacetime fleets. Surveying this locality was not, however, an easy job.

Tumbled into command of the *Beagle* at the age of twenty-three, FitzRoy had excelled. He restored the crew's collapsed morale by leading them in singing 'merry' sea shanties. He surveyed the difficult coastline with exemplary accuracy, dealt firmly but responsibly with thefts and attacks by natives, weathered a freakish storm that killed two men and careened the *Beagle* to the point of capsize, and won the admiration of his commander, Captain King, who'd originally wanted to appoint someone else to the position. It was not only FitzRoy's superb seamanship that garnered such respect, he'd also shown initiative and high moral aspiration by deciding to bring home four Fuegian hostages to be educated in the Christian religion at his own expense, with the aim of returning them to the region as translators and envoys for future naval surveys.[13]

Even so, young FitzRoy's ambitions to return to South America had initially been blocked by penny-pinching naval officials. He was on the verge of financing his own small ship to take the three surviving and freshly indoctrinated Fuegians back to their homeland, when the Lords of the Admiralty suddenly offered him command of the *Beagle* to undertake a new scientific survey to South America and the East Indies. They had been influenced by pressure from FitzRoy's uncle, Lord Grafton, and still more by the internal machinations of two Admiralty employees, who saw the young naval captain as an ideal figure to further their political ambitions.

John Barrow, Second Secretary of the Admiralty Board, was a former African explorer and a wily civil servant with indefatigable powers of persuasion. His patriotic zeal and love of geographical exploration fostered several generations of would-be naturalists. An incorrigible schemer who wrote regularly for the influential

*Quarterly Review*, he urged the Admiralty to deploy Britain's demobbed naval officers – currently unemployed on half-pay – as a peacetime force for extending Britain's imperial power. They would use their naval expertise to 'discover' unexplored lands, to map and chart sea passages, and to open up trade routes and settlements in New World regions like South America, Africa, Australia, and the Arctic and Antarctic circles.[14]

A key ally in this enterprise was Captain Francis Beaufort, the newly appointed Admiralty Hydrographer, a naval veteran who believed that the maritime sciences of hydrography, magnetic and astronomical charting, and weather prediction could revolutionise the navy and, in the process, extend Britain's geopolitical and economic reach.

FitzRoy was an ideal protégé. The well-connected young captain's expertise in naval sciences would instigate a new kind of surveying that made use of the latest scientific knowledge and precision technologies. He would help generate a scientific girdle for Britannia, a chain of accurate measurements that circled the world.[15]

This influential duo helped to prise open the tight coffers of the Naval Board, responsible for fitting out and supplying survey voyages. As a result the *Beagle* was refitted at a cost almost equal to her initial commissioning. She'd originally been a two-masted, ten-gun brig, of a type known wryly in the service as the 'coffin' class, because the low position in the water made the brigs vulnerable to capsize and swamping by even moderate waves. For surveying purposes, the *Beagle* had been converted to a barque. An extra mizzenmast at the rear, and the addition of cross-rigged sails at the bow and stern were designed to enable her to point closer into the wind.

Not satisfied, FitzRoy had her main deck raised, forward and aft, to give additional space below and to increase her stability. He had a new layer of fir planking and copper sheathing laid on the hull to improve her strength, speed and tonnage. To combat the legendary gales of Cape Horn, he also demanded the strongest spars and rigging the navy could procure.

If essential requests were refused, FitzRoy paid for them himself, including the provision of two extra whaleboats for navigating bays and rivers, and six of the total twenty-two chronometers to ensure exact navigational readings. As well as covering the cost of transporting the three Fuegians, and an English missionary to capitalise on their conversions, he financed an expert clockmaker to tend the chronometers, and an artist, Augustus Earle, to produce a visual record of the voyage. Finally he approached Beaufort for permission to carry a gentleman naturalist, who would be self-financing but supported on naval stores.[16]

FitzRoy's request for a supernumerary naturalist was unusual. The Admiralty and Naval boards were not opposed to having a naturalist on survey voyages, but insisted that naturalism should be an extra and unpaid part of the work of someone on the ship's muster. By convention, the position had come to be regarded as the province of the surgeon or assistant surgeon. The *Beagle*'s appointed surgeon, Robert McCormick, seemed appropriate. He was an experienced voyager and had ambitions of publishing a natural-history journal and building a reputation as a geologist.

What FitzRoy really wanted, however, was a gentleman companion with whom he could discuss diverting topics like philosophy, and with whom he could, if necessary, share more personal issues. He knew how isolated naval captains could feel on long survey voyages, a result of the need to display a sovereign-like authority over officers and crew. Having to exercise unquestioned power, enforced by corporal punishment, a captain could not afford to become intimate with his officers, however much he might like or respect them. Every naval commander knew how the soured friendship between Captain Bligh and Fletcher Christian had led to the mutiny on the *Bounty* in 1789.

FitzRoy believed he had a more than usually urgent need for such a confidante. He'd been stung by the grisly suicide of Pringle Stokes, whose life had leaked away in the *Beagle* captain's cabin over four days. The incident revived a long-suppressed fear in the young captain that he carried a hereditary taint of madness. A decade earlier, his famous uncle, Viscount Castlereagh, had slashed his throat with a cut-throat razor after developing an obsession that he was being blackmailed for a homosexual offence in Hyde Park. Already FitzRoy had felt premonitions of Castlereagh's fate in his own tendencies to explosive rage, his paranoid suspicion of colleagues, and his bouts of melancholy. There were times when he could hardly bring himself to speak or act. If he was going to survive the voyage of the *Beagle*, he needed to take a human antidepressant with him.

Francis Beaufort filtered FitzRoy's request for a gentleman naturalist through a variety of his Cambridge connections, two of whom, including Henslow, eventually turned down the offer for domestic reasons. Once Henslow realised that the captain's prime concern was for a companion, he thought of his personable young friend Charles Darwin. Genteel, congenial and domestically unencumbered, he seemed a perfect candidate.

Henslow's letter to Charles of 24 August 1831 conveying the offer was tactful and reassuring. He stressed that his recommendation was being made 'not on the supposition of yr being a *finished* Naturalist, but as amply qualified for collecting, observing & noting anything new to be noted in Natural History'. He also emphasised the importance of Darwin's social credentials: 'Captain F. wants a man (I understand) more as a companion than a mere collector & would not take anyone however good a Naturalist who was not recommended to him likewise as a *gentleman*.'[17] Without a second thought, Charles decided that he wanted to take up the offer.

Yet he was not completely confident he could do the job. Judging from his autobiography, written many years later, he shared some of his father's doubts. Charles acknowledged that much of his boyhood passion for collecting natural-history objects had been indiscriminate, and that his love of rambling, stalking and hunting was motivated largely by Crusoe-like fantasies. Later still, recalled Joseph Hooker, Darwin admitted that up until 1830 'the only objects of natural history . . . he cared for were foxes and partridges'.[18] Charles's own summation of his academic career was even harsher than his father's: 'during the three years which I spent at Cambridge my time was wasted . . . as completely as at Edinburgh and school'.[19]

He'd either squandered or failed to take up many opportunities to gain a serious grounding in natural-history subjects such as zoology and geology, and he cursed his lazy abandonment of the useful skills of dissection and anatomical drawing.[20] Despite becoming friendly for a time with the brilliant Edinburgh doctor and marine expert Robert Grant, Darwin had shown no interest in Grant's heterodox French ideas about the capacity of species to transmute by passing on useful adaptive traits to their offspring.[21] On Darwin's recent field trip around Snowdonia with Professor Adam Sedgwick, his geological knowledge had proved rudimentary. He simply didn't know what to look for. After a short time, he'd left the professor searching for fossils on the shores of a small lake at Cwym Idwal to launch out on a private hike through 'some strange wild places' in the mountains, before eventually joining student friends at the seaside town of Barmouth.[22]

Even Darwin's friendship with Henslow, which had inspired him to become a more serious student of nature during his last terms at Cambridge, was fuelled as much by his admiration of the professor's moral and poetic qualities. The energetic, outdoor Henslow was the kind of muscular clergyman Charles wanted to be. Most of all, the two came to share a romantic dream of exploring tropical places. Henslow had nursed this fantasy ever since winning as a school prize a copy of François Levaillant's lively travel book on mysterious Africa.[23] He in turn infected his young protégé with 'tropical fever' by lending him the Prussian scientist Alexander von Humboldt's *Personal Narrative of a Journey to the Equinoctial Regions of the New Continent*. This was a huge, sprawling account of a scientific exploration of Brazil with companion Aime Bonpland from 1799 to 1804. The flowery English translation by Romantic English writer Helen Maria Williams had the effect of exciting Darwin like nothing he'd read before.[24]

For much of 1830–31 he could think of little else, even though he was attracted at this time to a young family friend, Fanny Owen. Fanny aside, Charles became fixated with the idea of taking a tour of Tenerife, the subject of one of Humboldt's most evocative pieces of writing.[25] Darwin's imagination fizzed with images of palms, lianas, orange groves, volcanos and sandy beaches. He confessed in letters to his sisters that he deliberately stoked his 'Canary ardour' and worked himself into a 'tropical glow' by visiting hothouses and zoos, his copy of Humboldt in hand. At every opportunity he read out passages to friends, and badgered them to accompany him on a short collecting tour.[26]

Only by visiting these places, Darwin maintained, could they hope to experience them in the same intense way Humboldt had. He wanted to feel the same riot of emotions, the same intoxicating rush of speculative inspiration. 'I never will be easy,' he vowed to his sister Caroline, 'till I see the peak of Teneriffe [*sic*] and the great Dragon tree; sandy, dazzling plains, and gloomy silent forest are alternately uppermost in my mind.'[27]

Dr Darwin thought this Tenerife obsession more a romantic fantasy than a serious naturalist excursion – a last hurrah of youth. Charles, for his part, knew that he possessed capacities, latent and newly aroused, that had impressed accomplished Cambridge dons like Grant, Henslow, Sedgwick and Whewell. He knew that his curiosity about nature was limitless, that he could work on things of interest to him as intensively as any scholar, that he was prepared to tackle difficult problems and to speculate freely about their solution. He confessed to Henslow in July 1831, 'As yet I have only indulged in hypotheses; but they are powerful ones, that I suppose, if they were put into action but for one day, the world would come to an end.'[28]

Some of the attributes that Dr Darwin judged frivolous Charles knew to be valuable. He was strong and fit, a good rider and a keen shot – all essential skills for collecting geological, faunal and floral specimens. The hunter-collector in him was based on a fascination with the habitats and behaviours of birds and animals. If he wasn't a 'finished naturalist', he had at least laid foundations. He'd learnt from his boyhood beetle-collecting how to observe, capture, preserve and display specimens; from the ex-slave John Edmondstone how to stuff birds; from Grant how to use a microscope to uncover the minute internal structures of marine species; from Jameson the elements of zoological anatomy and classification; from Henslow how to identify the structures, taxonomies and locations of rare plants; and from Sedgwick how to read basic stratigraphy and take samples. Charles Darwin was no dunce.

Henslow's prescribed reading for the Tenerife trip had also taken Darwin well beyond mere romance. Poring over Humboldt's *Personal Narrative* in 1831 taught him that the aesthetic pleasures of the picturesque traveller and the intellectual satisfactions of the empirical scientist could be complementary. Humboldt believed that the naturalist should seek to capture the unity and harmony of nature in all its diversity, and hence that poetic and painterly abilities were as important as empirical and mathematical ones.

This chimed with Darwin's own aesthetic inclinations, which

were informed by a liking for English poets, such as Wordsworth, Coleridge and Shelley, as well as a knowledge of contemporary art theories of the picturesque and the sublime. Along with the Lake poets, he believed that the intuitive 'inner sense' of a cultivated, sensitive mind might at a glance grasp the harmony and unity that underlay nature's physical and organic diversity – what Wordsworth in *Lyrical Ballads* (1798) described as 'the passions of men . . . incorporated with beautiful and permanent forms of nature'.

Humboldt also taught him that the naturalist should train himself to measure the precise material characteristics of climate, soil, humidity, elevation and species that shaped the environments of distinct natural regions. The data obtained should be used comparatively so as to devise a picture of the whole of nature.

At the same time, Henslow's recommendation of a new book by the British astronomer Sir John Herschel, *A Preliminary Discourse on the Study of Natural Philosophy*, excited Charles with the sweeping potency of the scientist's methods. Attention to true cause and empirical facts would help uncover objective laws. For the first time Darwin felt 'a burning zeal' to make some mark in science, in the manner of these two great explorers of the natural world.[29]

Kindly Dr Darwin could hardly fail to notice his son's palpable disappointment at having to decline Henslow's offer to travel around the world, and he suddenly made a critical concession. If any man of good sense could be persuaded to endorse this harebrained scheme, he told his son, he was prepared to reconsider his opposition.

As he probably intended, this goaded Charles into riding to Maer to recruit the support of his industrialist uncle, Josiah Wedgwood. Sure enough, all the Wedgwood family were sympathetic to the idea of the voyage, and anxious that the doctor revise his judgement. Charles accompanied Josiah Wedgwood in his carriage back to

Shrewsbury, fretting all the thirty miles. Once there, Uncle Jos proceeded to lay out an astutely modulated rebuttal of all Dr Darwin's objections.

Though naturalism was not a profession, Uncle Jos conceded, it would surely do Charles no harm as a clergyman and a gentleman to have his intellectual horizons widened in this way. The fact that others more qualified had declined the position was no reflection on Charles, or on the calibre of the naval expedition: he 'could not conceive that Admiralty would send out a bad vessel on such a service'. On the crucial question of whether the voyage might weaken Charles's resolve to settle down as a clergyman, Uncle Jos presented a telling point: 'You are a much better judge of Charles's character than I can be. If on comparing this mode of spending the next two years with the way in which he will probably spend them if he does not accept this offer, you think him more likely to be rendered [un]steady, and unable to settle, it is undoubtedly a weighty objection. Is it not the case that sailors are prone to settle in domestic and quiet habits?'[30]

It was a decisive thrust, not least because long-suffering Dr Darwin had just received a fresh swag of Charles's unpaid Cambridge bills. Sensing victory, Charles chipped in with the cheeky consolation that 'I should be deuced clever to spend more than my allowance whilst on board the *Beagle*'. 'Yes,' the doctor replied. 'But they tell me you are very clever.'[31]

It remained for Charles to convince FitzRoy to accept him formally as the *Beagle*'s naturalist. This proved more difficult than expected because the captain was beginning to have doubts about the whole idea. Perhaps his concerns were prompted by Beaufort's rather extravagant description of Darwin as 'a savant', or perhaps FitzRoy, a deep-dyed Tory, had heard from somewhere that his prospective companion was a damned Whig. As likely as not, the captain simply wondered about the wisdom of having an unknown person living with him at such close quarters. What if their personalities grated? The inescapable proximity of such a person could actually worsen his depressive tendencies. Furthermore, the presence of the Fuegians and other supernumeraries now meant that the *Beagle*'s muster was stretching the accommodation beyond its limits.

On their first meeting, FitzRoy therefore set himself to be discouraging. He voiced his doubts about available space on the ship, and indicated that he'd already invited a distant cousin to serve in the position. Did Mr Darwin realise, moreover, that they were unlikely to be going to the South Seas but rather to Patagonia, one of the harshest environments in the world? The voyage could be protracted beyond two years and would anyway be full of hardship and danger. It was certainly not a prospect for faint hearts. FitzRoy later admitted that he'd also made a snap phrenological diagnosis that the shape of Darwin's nose suggested irresolution.

But in spite of his squashy nose and reputation as a savant, Darwin won FitzRoy over. His strong physique, easy personality and gentlemanly pedigree dissolved all doubts. The cousin vanished from consideration and the two young men discussed guns and shooting, nautical and natural science, and the pleasures of southern landscapes. Darwin's excitement was infectious, his deference pleasing. FitzRoy, now convinced, set about finding his new companion even better accommodation than he'd expected.

Darwin meanwhile returned to his family raving about the handsome, intelligent and gentlemanly sailor, so self-assured and experienced that he was reminiscent of Nelson. He scribbled a note to Henslow on 5 September 1831: 'What changes I have had: till one today I was building castles in the air about hunting foxes in Shropshire, now Llamas in America. – There is indeed a tide in the affairs of men.'[32]

His excited sisters nicknamed FitzRoy 'Captain Wentworth', after

Jane Austen's romantic hero, but an offer to provide the captain with a copy of *Persuasion* was rejected.[33] Even so, with his tendency to hero-worship, Darwin declared the slight, dark and handsome FitzRoy his 'beau ideal of a captain'.[34]

Darwin didn't expect, though, to have his fortitude tested even before embarkation, preparation for which proved more protracted than anyone anticipated. Charles and his father worried initially that there wasn't enough time to make ready for the voyage: the letter of offer had suggested the *Beagle* would sail within a month. September disappeared in frenzied activity, with Dr Darwin proving characteristically generous now that he'd committed himself to releasing his son to this wild adventure. Charles bought clothes, books and equipment for the trip, including pistols and a fine rifle, a bargain that he assured his father might easily have cost twice as much. As he chortled to one of his friends, he'd have need of such weaponry, 'for we shall have plenty of fighting with those d—d Cannibals.'[35]

He consulted experts on the Southern Hemisphere, including the surveyor Captain King, who talked to him about meteorology and whose young midshipman son would share Darwin's tiny poop cabin. When inspecting his cramped sleeping space, he met another of its future occupants, the mate John Lort Stokes, who was to draw up his charts on the same small table where Darwin would pore over his microscope. Robert Brown, the shy scholar at the British Museum who'd circumnavigated Australia with Flinders and become the world's authority on Australian flora and fauna, showed Charles a range of Antipodean specimens.

On 8 November, Darwin moved his things aboard, but it was almost another month before he began sleeping on the ship. For most of this time he found himself half numbed, half awed by the frantic pace at which the *Beagle* was fitted out and victualled. He confessed to his sister Susan, 'My spirits about the voyage are like the tide, which runs one way and that is in favour of it, but so by a number of little waves, which may represent all the doubts and hopes that are continually changing my mind.'[36] Hesitantly he began a journal, as much as anything to give himself some sense of purpose while a bewildering world sprang into shape around him.

In the event, delays ensured that there was ample time for him to discover just how alien this new wooden world would be. Thanks to extreme adverse weather conditions, the *Beagle*'s attempts to set sail had twice to be aborted. While docked in Plymouth in November and December, Charles missed his family with an aching homesickness that he'd never before experienced. He learnt how awkward it was to get into a narrow naval hammock slung in a space so small that he had to remove drawers from a chest each night to make room for his six-foot frame – he slept with his head protruding into the space thus created. He experienced the disturbing insularity of the ship's officers: when not talking obsessively about sail settings in a language he found as unintelligible as Hebrew, they enjoyed alarming him with tall tales about the rituals that Neptune would inflict on him when they crossed the equator.

He began to discover, too, that like most people with a little medical knowledge, he was a hypochondriac. Racked with stomach pains and heart palpitations, he wondered whether he might have some terrible disease, though part of him sensed that worry about his ability to cope might lie behind it all. He urged himself in his journal to develop steady and industrious methods of work and not throw away 'an opportunity of improving myself'.[37]

Worst of all, he had time during their short attempts to beat out of the harbour in battering seas to understand just how susceptible he was to seasickness. 'I suffered most dreadfully,' he wrote on 7 December as they doubled the breakwater, 'such a night I never passed, on every side nothing but misery; such a whistling of wind and roar of the sea, the hoarse screams of the officers and shouts of the men, made a concert I shall not soon forget.'[38] The ship was 'full of grumblers', but he had no doubt that he was the worst, openly doubting whether the voyage would ever add to the happiness of his life.

FitzRoy did his best to help. As Caroline wrote in reply to her brother's woeful complaints, 'Papa's eyes were full of tears when he thought of your first miserable night and then of your good natured Captain, in all the confusion, paying you a visit & arranging your hammock:– . . . everything you tell us of him makes him more and more perfect.'[39]

Just before they finally managed to leave Britain's shores on 27 December, Darwin had a glimpse of insight into why his father had been so doubtful about the social respectability of sailors. On Christmas and Boxing days, most of the crew managed to get themselves drunk with an animality that shocked even someone familiar with student binges. Darwin conceded the necessity of the severe discipline that followed, but it sickened him. He watched miserable sailors in irons crying and abusing each other, and for days could not get the sounds of their floggings out of his mind.[40]

All in all, he later recalled, those two months of false embarkation were 'the most miserable I ever spent'.[41] His normally cheerful personality was swamped by 'dark and gloomy thoughts'.[42] More than once he replayed in his mind the portentous conversation with his father at the Mount on 29 August 1831, and wondered whether Dr Darwin might just have been right about what a harebrained idea it was for his son to voyage around the world.

# The Philosopher at Sea

Early in the voyage, FitzRoy gave Charles Darwin the nickname of Philos, short for Philosopher. It was a sign of the captain's growing ease with the young gentleman he had riskily selected to help maintain his sanity. The officers quickly took it up as well, but it's not clear what the name signified for them. The word 'philosopher' at that time referred to someone devoted to the study of such subjects as the character of man, nature, morality, literature and art.

At sea, however, being called a philosopher had an edgier connotation. A philosopher was not as bad as a Jonah – there is plenty of evidence of Darwin's popularity among both officers and seamen – even so, it was a sobriquet that mocked its owner as a creature out of his element.

Charles Darwin admitted to having boarded the *Beagle* in 1831 thinking of a ship as simply 'a large cavity containing air, water and food mingled in hopeless confusion'.[1] Within weeks he was complaining that even the simplest shore-based comforts, such as washing oneself with soap, became 'so much extra trouble at sea' that they were scarcely worth doing. A month out of harbour he exploded in exasperation to his diary, 'Oh a ship is a true pandemonium!'[2] He puzzled at the obsession with naval rank that made the conversations of his officer colleagues in the gunroom mess 'so devoid of interest'.[3] The relentless misery of seasickness made him curse both the sailor's vehicle and its element: 'I hate every wave of the ocean with a fervour,' he told his cousin. 'I loathe, I abhor the sea and all ships which are on it,' he confessed to his family.[4]

The mutual incomprehension between a landlubber philosopher and a ship's crew of this era is brilliantly realised in the fiction of Patrick O'Brian, whose naturalist-surgeon character Stephen Maturin has a shipboard ignorance and dislike of naval discipline that provokes the wry tolerance of the sailors he tends.[5] Darwin's nickname was probably also a reflection of his ambiguous standing and role on the ship, since he had been given no formal naval appointment. The official *Beagle* naturalist was surgeon-geologist Robert McCormick, and, after him, assistant surgeon Benjamin Bynoe. Several other officers could also make scientific claims: FitzRoy in geology, Lieutenant Wickham in botany, Second Lieutenant Sulivan in geology; even fourteen-year-old midshipman Philip Gidley King had learnt some zoology and botany from his surveyor-botanist father. No wonder the raw young Darwin was not at first taken seriously: he was amiable Philos, the captain's gentleman entertainer.

It's not clear how Darwin himself viewed his status as he struggled to adjust to the unsettling motion and routine of his 'floating prison'.[6] There's no indication that he minded being called Philos, or sensed any condescension. More likely he welcomed the nickname as a sign of his sovereign captain's approval: he'd initially felt 'overwhelmed' by the variety of subjects he would be expected to discuss at FitzRoy's table.[7]

A philosopher was probably a pretty fair description, too, of his self-image at the beginning of the voyage, not least because it fitted the wide-ranging accomplishments of his hero Alexander von Humboldt, whose descriptions of tropical landscapes had made the prospect of a southern voyage so appealing.[8] In his later years Darwin was to say that the whole course of his life was due to having, as a young man, read and reread Humboldt's *Personal Narrative*.[9] Henslow had given him an inscribed copy before he set off, and he read it over and over. Lying in his hammock or on the captain's sofa clasping *Personal Narrative* and a handful of raisins seemed the only thing that could ease his queasiness. It was better 'than even novels' for vaulting him out of the miseries of his swaying hammock and onto the solidity of island beaches and Brazilian forest floors, triggering in his mind the sensual pleasure first experienced when reading of Tenerife's tropical scents and luminous colours.

During those early months at sea, Darwin chafed to leave the ship and replicate Humboldt's experiences. He wanted to conjure up the same rapturous emotions and to transfer them poetically to others – to his family and friends, of course, but also perhaps to unknown future readers. In May 1832, he ruminated on the mysterious sympathetic power of a great writer's prose: 'Few things give me as much pleasure as reading the *Personal Narrative*. I know not the reason why a thought which has passed through the mind, when we see it embodied in words, immediately assumes a more substantial and true air. In the same manner as, when we meet in dramatick writings a character we have known in life, it never fails to give pleasure.'[10]

Darwin had begun to discover the pleasures of writing a daily journal soon after moving onto the ship in Plymouth harbour. It was one way to impose order on the chaos around him during the frenetic period of the *Beagle*'s refitting and victualling. Behind his daily entries lay also a vague ambition to emulate Humboldt's naturalist journal. Darwin had packed among his books a copy of Edinburgh flower painter Patrick Syme's edition of *Werner's Nomenclature of Colours*. This artist's handbook enabled the reader to match colour samples to technical descriptors, a handy tool for a would-be writer.[11]

No wonder, then, that he'd felt such anguish in January 1832 when the *Beagle* bypassed the port of Madeira because of rough seas, then had to leave the tropical paradise of Tenerife without landing because of cholera quarantine restrictions. 'Oh misery, oh misery,' he mourned in his diary. From the deck, the island appeared to be every bit the idyll Humboldt had described. Darwin could see gaudy red and yellow houses, ornate oriental churches, and banks of trading vessels with raked masts moored against 'the magnificent background of Volcanic Rock'. The air felt 'deliciously warm' and he could hear the sound of waves rippling against the stern and see the reflection of stars shining on the water, as bright as 'little moons'. It reminded him of paintings of the gods presiding in Olympus.[12]

Perhaps the double deferral was a good thing: it intensified Darwin's hunger for tropical habitats and gave him time to get used to some of the shipboard routines.

A fortnight later, on 16 January 1832, the *Beagle* anchored at Porto Praya, chief port of St Jago (Sao Tiago) in the Cape Verde Islands, some 435 miles off the coast of Africa. The harbour's rather desolate look vanished immediately Darwin landed, to be replaced by 'the unspeakable pleasure of walking under a tropical sun on a wild and desert island'. He revelled in 'the glory of tropical vegetation' as he strolled among groves of tamarinds, bananas, coconuts and palms. He wrote ecstatically:

> I expected a good deal, for I had read Humboldt's descriptions, and I was afraid of disappointment: how utterly vain such fear is, none can tell but those who have experienced what I today

> have . . . the numberless and confused associations that rush together on the mind, and produce the effect. I returned to the shore, treading on Volcanic rocks, hearing the notes of unknown birds, and seeing new insects fluttering about still newer flowers. It has been for me a glorious day, like giving a blind man eyes, he is overwhelmed with what he sees and cannot justly comprehend it. Such are my feelings . . . [13]

The moods of the tropics vibrated in Darwin's soul. His future friend Joseph Hooker was to write that 'nature in all its aspects spoke to [Darwin's] feelings with a voice that was living and direct'.[14] Like the Romantic poets, whom he read almost as often as he read Milton, Darwin's emotions were threaded with nostalgia. He mourned the passing of a moment of rapture, yet thrilled at the 'the pleasure . . . [of] anticipating a time when I shall be able to look back on past events'.[15]

At the moment when the *Beagle* passed across the equator into the Southern Hemisphere on 17 February 1832, Darwin had an epiphany. After recovering from the sailor's ritual of being shaved by a sailor dressed as Neptune with a piece of blunt hoop-iron, he reflected, 'in quiet solitude', on how far he'd travelled from home, both geographically and psychologically. Gazing at 'the bright band of stars 'from Orion to the Southern Cross', he experienced 'a kindred feeling' across time and space with his predecessor Humboldt, who'd similarly looked up at the Southern Cross and felt 'one of the dreams of my earliest youth come true'.[16]

Darwin found it difficult at first to make sense of scenery so novel that it 'bewilders the mind'. Towards the end of February, he was again overwhelmed by 'transports of pleasure' when the *Beagle* put in at the Bay of All Saints in Bahia (St Salvador), on the east coast of modern Argentina. Tramping through his first rainforest reminded him of 'the Arabian Nights, with the advantage of reality'.[17]

A month later, he stayed for two days on a large estate, inland from Rio de Janeiro. The beauty of the forest confronted him with one of the Romantic artist's stock dilemmas: 'it is easy to specify individual objects of admiration; but it is nearly impossible to give an adequate idea of the higher feelings which are excited: wonder, astonishment and sublime devotion, fill and elevate the mind'.[18]

FitzRoy chuckled that his young friend Philos seemed like a child with a new toy.

Yet Charles Darwin was much more than a star-struck Romantic: he wanted to match Humboldt's 'rare union of poetry with science'. Following his hero, Darwin believed that the naturalist's task was to discover both the diversity and the underlying unity and harmony of nature. Like a poet, he would work to penetrate the mystery behind the veil of reason, but he would do so in the service of science. This mission meant also using physical instruments, such as barometers, and mathematical techniques, such as statistics, to uncover the laws of nature. Ideally, the naturalist should aim to collect, classify, measure and map the whole natural order.[19]

To aid in this vast undertaking, Humboldt had divided the natural world into a series of *divisions physiognomiques*, regions shaped by local climates and soils. These regions could be identified by the presence and distribution of representative plants, such as firs, cacti and grasses, which had developed distinctive interrelationships with other flora and fauna within their habitats. The task of the Humboldtian naturalist was to grasp the environmental character of an individual region and then compare this with other regions, until the 'natural economy' of the world could be understood. In short, Humboldt envisaged an ecological mapping of the entire globe.[20]

Darwin set about this challenging task with the same vigour he'd shown hunting and collecting beetles as a boy. He gathered plants, birds, rocks and marine creatures, using his gun, pick, jars, and a dredging net that trawled daily behind the ship. He also had specialised instruments – a clinometer for determining rock gradients, a hygrometer for measuring air moisture, and a microscope for observing the inner structures of specimens.

His zeal was not always appreciated. The sight of Darwin's mounting collections, and his use of the naval postage system to send specimens to Henslow in Cambridge, so infuriated the official naturalist, McCormick, that he left the ship at Rio, cursing FitzRoy for allowing an unqualified outsider to usurp his domain. Darwin thought it a good riddance: the man was a pompous ass with antiquated scientific ideas.[21]

Darwin applied a new regime of self-discipline when describing, assessing, cataloguing, preserving and recording his findings. He'd been introduced to the idea of scientific exactness in Cambridge, but the shipboard rituals of navigating, mapping, logging, sketching, surveying, depth-sounding, and magnetic and meteorological observation gave him a new 'habit of energetic industry and of concentrated attention to whatever I was engaged in'.[22] This 'training' or 'drilling my mind', as he called it, began even before the ship departed. His cabin-mate John Lort Stokes taught him how to make magnetic readings with a dipping needle, Captain King instructed him in meteorology, and FitzRoy showed him how to take latitude by measuring the angle of the midday sun, and longitude by calculating the time difference between local midday, with the sun at its zenith, and Greenwich time as registered on the ship's chronometers.

'I find to my great surprise,' Darwin wrote to his father on 1 March 1832, 'that a ship is singularly comfortable for all sorts of work. – Everything is so close at hand and being cramped makes one so methodical, that in the end I have been a great gainer.'[23] If he needed reference books, the library contained some two hundred and fifty items, not insignificant by any scholar's standards, and his ever-tolerant father supported a further stream of purchases throughout the voyage. For Charles Darwin, the *Beagle* was a type of college, with multidisciplinary knowledge available in an instant, including the artistic talents of landscape painter, Augustus Earle, and his later replacement Conrad Martens.[24] Darwin thus became habituated to the post-noon rhythms of shipboard work: while FitzRoy filed his daily logs and made surveying calculations, Darwin would write his journal, transcribe information from notebooks, or tag, number and describe his latest specimens.[25]

The post-voyage differences that opened up between Darwin and FitzRoy, whose religious fundamentalism intensified after his marriage, have led people to underestimate the depth of their friendship while on board, the similarity of their scientific ideas, and the degree of reliance Darwin placed on his captain and dinner companion. True, they clashed on political and social issues at times, most famously when Darwin's liberal-emancipist views recoiled from FitzRoy's Tory-paternalist support for Brazilian slavery. For most of the voyage, however, their ideas on natural history, natural philosophy and even on religion were remarkably compatible and reinforcing. The image of a bible-thumping FitzRoy emerged after the *Beagle*'s return to England.

For the bulk, if not the whole, of the voyage, Darwin and FitzRoy both believed – like many other British naturalists of the day – that the earth had existed for much longer than the six thousand years claimed in the Bible. They both also believed that species were divinely created and were fixed and permanent, but capable of developing small, temporary variations as a result of environmental shocks and adaptations. They assumed that species had probably been brought into being on earth by more than one divine creation over time, and that such creations had taken place within special 'centres', from which plants and animals occasionally migrated in response to changes in geological and environmental conditions. They further held that at the time of creation, each species was perfectly adapted, or designed, to suit its environment. And in accordance with the divine plan, each fresh creation over time, including that of humans, revealed greater complexity and perfection.

At the start of the voyage, both Darwin and FitzRoy saw the natural order as relatively static and harmonious, though interrupted by catastrophic events of the past, such as the Flood, which had brought about extinctions. However, this belief in a stable globe was – as we shall see – modified during the trip by their common reading of Charles Lyell's *Principles of Geology*. Both men also agreed that the primary task of the naturalist was to collect, describe and classify as many as possible of the objects of nature so as to show the complete arrangements or 'orders' of species.[26]

And if FitzRoy was more liberal in his views on the natural world than is usually thought, Darwin became gradually more sympathetic to the ways of ships and sailors than some biographers have allowed. Unlike Patrick O'Brian's sarcastic philosopher-surgeon Maturin, who remained obstinately resistant to naval culture, Darwin eventually found much to admire in shipboard life – at least during the periods when he was not too seasick. His diary shows that an early flood of 'naval fervour' was strengthened by a gradual adaptation to both the 'quiet and regularity' of ship routine, and to the busy hum of action when sailing in brisk wind on the open ocean.[27]

From the outset, Darwin was impressed by the machine-like efficiency of the sailing ship, especially the 'rapidity and decision of the orders and the alertness with which they were obeyed'.[28] He could not have been more proud on 4 April 1832, when the *Beagle*'s crew showed off their sail-handling skills in Rio harbour. 'It was quite glorious today how we beat *Samarang* in furling sails,' he boasted in a letter to his sisters. 'It is quite a new thing for a sounding ship to beat a regular man o' war.' Young Philip Gidley King recalled that Darwin had held 'a main-royal sheet in each hand and a top-most studding-tack sheet in his teeth', and had afterwards claimed that 'the feat could not have been performed without him'.[29] Darwin told his family that he was becoming such an accepted sailor that he even ventured at times to sit down 'in the sacred precincts of the quarter deck'.[30]

A month later, he admitted to some envy when watching a 74-gun, line-of-battle ship, the *Warspite*, bristling with four hundred men in white, practise armed manoeuvres in front of the Admiral.[31] The experience led him to spend the whole of the following day imagining 'the pomp and circumstance of war'.

As it happened, the daydream turned to reality. In early August 1832, the governor of Montevideo, a fortress town on the River Plata in modern Uruguay, asked FitzRoy to help quell a local uprising by some mulatto rebels. Fizzing with excitement, Darwin joined the invading force of fifty-two sailors and marines, all armed with muskets and cutlasses. The piratical-looking crew were then loaded into the *Beagle*'s four small boats to attack the central fort where the insurgents were located.

Philos was itching for a fight. He told his sister Susan proudly that 'it was something new to me to walk with Pistols and Cutlass through the streets of a Town. – [but] it has all ended in smoke'.[32] To his chagrin, the rebels surrendered without firing a shot. Even so, he marvelled at the 'reckless gaiety with which sailors undertake even the most hazardous of attacks'.[33] This, too, could be seen as another successful rite of passage.

In the aftermath of the phantom mêlée, Darwin reflected on how different a person he had become from the naïve boy who'd arrived back home from hiking in Wales less than a year before to find the offer of the *Beagle* voyage awaiting him.[34]

One major difference between then and now was that Charles Darwin had become a seaman. In spite of his proneness to seasickness, he felt the sailor's thrill of defying some of the biggest seas in the world. On 10 July 1832, he was feeling sufficiently well to stand on deck as they sailed into gale-strength winds off Cape Horn. Even though 'the sky looked very dirty and the waves with their white crests dashed angrily against the ship's sides', he thought it 'a beautiful spectacle to see how gracefully the *Beagle* glided over the waves, appearing as if by her own choice she avoided the heavy shocks'.[35] His diary entry of 22 December 1832 flaunted a seaman's terminology when recording that the *Beagle* had doubled Cape Horn, close-reefed in a storm, yet made remarkably little leeway. Just over a week later, they weathered the ugly rocks off Diego Ramirez in similarly heavy conditions, leading Darwin to boast that, despite being close-hauled with topsails and courses reefed, the *Beagle* was logged at a sprightly seven and a half knots.[36]

But on 13–14 January 1833, they ran into a Cape Horn storm that even the unflappable FitzRoy was to call the worst of his life. At one point the *Beagle* was struck sideways-on by a wave taller than the topmast. The little 'coffin ship' careened so severely that 'sea filled the decks', and the stern davits plunged three feet under water. Sailors struggled to cut away a flooded whaleboat that was driving the ship's gunwales further under, until the quick-thinking carpenter knocked open the ports, allowing water to pour out and the ship to right itself. Another such wave, FitzRoy said, would have sunk them. Darwin grumbled about the water damage to his plants and papers, but was actually so proud of the resilience of ship and crew that he began to call the *Beagle* 'our little diving duck'.[37]

Despite his growing confidence at sea, however, crippling seasickness, a thirst for adventure, and the needs of natural history drove Darwin to grab every chance to undertake expeditions ashore. As a result, he spent three-fifths of the overall voyage on land.

It was here that he showed his daring. Whether walking or riding on horseback, he thought of himself as 'a wanderer', eager to follow adventures as they came. Love of the chase – 'an instinctive passion' – made him feel like a hunter again. Looking back on the voyage in the closing pages of his diary, he concluded that these overland explorations had fulfilled a primal longing: 'the pleasure of living in the open air, with the sky for a roof, and the ground for a table . . . It is the savage returning to his wild and native habitats.'[38]

The *Beagle*'s arrival in Argentine frontier country in September 1832 soon found Darwin 'shooting, riding, collecting and looking forward to the thrill of a few revolutions'. He galloped on horseback across the grassy pampas and he learnt the difficult art of hurling the *bolas* – twin balls on a leather thong – to snag the legs of running prey. Naturalist he might be, but he claimed in his diary that he had never enjoyed himself so much as when hunting ostriches in the company of half-caste Argentinian-Indian soldiers.[39]

From August to early October of the following year, while the *Beagle* renewed its laborious surveying triangulations up and down the east coast of South America, Darwin was able to travel back and forth over the same terrain. This included trekking five hundred miles across wild Indian country with gaucho cattlemen as riding companions. Their black moustaches, long curling hair, proud faces and clanking spurs reminded him of desperados in a Salvator Rosa painting. Somehow, though, they managed to combine ferocity with the most polite manners.[40]

In August 1833, he briefly met the head of the army, General Rosas, whose gaucho dress and behaviour confirmed his reputation as a cruel soldier and hunter of Indians. Not to be cowed, Darwin and his servant Syms Covington rejoined the *Beagle* in October by sneaking through a Montevideo blockade set up by Rosas's soldiers.[41] Philos was clearly someone to be reckoned with.

If Darwin's courage helped erode his image as a landlubber, growing self-confidence reinforced his status as a naturalist. In September 1832, FitzRoy's sharp eyes detected fossil bones protruding from high up in the Punta Alta cliff-side. Without hesitation, Darwin took control of the operation of hacking them out of the silt and spreading the chunks of soil-encrusted bone, teeth and tusk on the deck for cleaning, sorting and cataloguing. From an area of around a hundred and fifty square yards, he and the crew extracted 'a tolerably perfect head of a megatherium, and a fragment and teeth of two others . . . an animal of the order of Edentata, as large as a pony . . . with great scratching claws . . . two great Edentata . . . both fully as large as an ox or a horse . . . another large animal closely allied with [a] Toxodon . . . a large piece of the tessellated covering like that of the armadillo, but of gigantic size . . . a tusk . . . which closely resembles that of the African boar', as well as scores of fragments of shells of marine creatures.[42]

Smaller caches were also uncovered thirty miles away in another red soil cliff. Although the fossils would need to be examined back in London by an expert palaeontologist, such as Professor Richard Owen, Darwin knew that they were the remains of extinct animals, and guessed that some were 'closely similar to species now living in the same bay'.[43]

This last was a surprise. Darwin had assumed, along with many geologists, a fundamental discontinuity between the past and the present among fossil remains. Extinct animals supposedly bore no ancestral relation to living species. To assume otherwise would imply transmutation, a crazy idea espoused only by a few followers of the French revolutionary naturalist Jean-Baptiste Lamarck. Here was a mild conundrum for Darwin. He mulled it over as he steadily accumulated fossil samples. His reading also told him that similar puzzles relating to 'the law of succession of types' awaited him in Australia, 'where fossil remains of a large and extinct species of Kangaroo and other marsupial animals were discovered buried in a cave'.[44]

Supervision of this fossil excavation marked another rite of passage for Darwin. He was now the undisputed *Beagle* naturalist.

Henceforth fossils and other interesting specimens found by the crew came straight to him for evaluation.[45] Even before this, Darwin had begun to test his rising speculative powers in another field of natural history and philosophy, the study of the distribution and character of races and groups of peoples – what we would now call ethnography. His opportunity for this came through FitzRoy's undertaking to return the three now 'civilised' Fuegians, Jemmy Button, York Minster and Fuegia Basket. The captain hoped they would provide a vanguard for a new Patagonian mission, to be led by a young missionary, Mr Matthews.

The reintroduction of the Fuegians to their original tribal groupings, and the setting up of the mission, gave Darwin a perfect opportunity to think about the role of nature and nurture in shaping individual character, and of savagery and civilisation in influencing morality and behaviour. He noted that Jemmy Button must have given Matthews pause for thought by telling him that in the winter his fellow countrymen 'sometimes eat the women'.[46]

The *Beagle* arrived at Tierra del Fuego in mid-December 1832 and gradually moved northward, encountering stray groups of Fuegians as it sailed along the coast. The first sight of these 'wild' people shook Darwin profoundly. On Christmas Day he could barely contain his horror at the contrast between their three Fuegian passengers, who dressed and behaved for all the world like Britons, and a fellow tribeswoman without any clothing who was suckling a newborn child 'whilst the sleet fell and thawed on her naked bosom, and on the skin of her naked child'. He concluded that within this harshest region of South America, mankind 'exists in a lower state of improvement than in any other part of the world', though he later ranked some South African and Aboriginal Australian tribes not far behind. He assumed, however, that different stages of civilisation and savagery were a result of historical rather than biological factors. Civilisation remained a possibility for all humans.[47]

At the same time, he was not much surprised when the *Beagle* revisited Patagonia in April 1834 to find the missionary Matthews in fear of his life and Jemmy Button now 'quite naked, excepting a rag round his waist'. York and Fuegia had moved to their own part of the country, but not before York had stolen Jemmy's clothes.[48]

Compassion and condescension jostled each other as Darwin attempted to sum up the significance of what he had seen. 'These poor wretches,' he wrote in his *Voyage of the Beagle*, 'were stunted in their growth, their hideous faces bedaubed with white paint, their skins filthy and greasy, their hair entangled, their voices discordant, their gestures violent and without dignity. Viewing such men, one can hardly make oneself believe they are fellow creatures, and inhabitants of the same world.' He even wondered whether, 'like some of the less gifted animals', they were capable of such social pleasures as enjoying a home or expressing affection. In this arid and miserable place, he asked, 'what is there for imagination to picture, for reason to compare, for judgement to decide upon?'[49]

Still, he had the evidence that a relatively short period of education could transform these same savages into people who displayed all the morality, manners and mental powers of Britons. Humans, whatever their external appearances, had surely all been drawn from the same common stock. 'We must suppose,' he concluded on a more sympathetic note, 'that [the Fuegians] enjoy a sufficient share of happiness (of whatever kind it may be) to render life worth having. Nature by making habit omnipotent, and its effects hereditary, has fitted the Fuegian to the climate and productions of his country.' Henceforth it became one of his keenest satisfactions 'to behold the various countries and many races of mankind' and to speculate how they had acquired their particular cultures.[50]

Darwin's transformation from the slightly comical figure of Philos into a self-confident naturalist owed most to his absorption of a set of ideas from a book that FitzRoy had presented to him at the beginning of the voyage. This was the first volume of Charles Lyell's *Principles of Geology*, published the year before the *Beagle* set sail. During the months and years at sea, it gradually supplanted Humboldt's *Personal Narrative* as Darwin's chief inspiration and text-book. Written by a former barrister turned naturalist, *Principles* was a scientific work unlike anything Darwin had read before. Fluent, tautly argued and empirical, with a global sweep and a deep grasp of contemporary European geological literature, it gave Darwin a new field of professional mastery, a new scientific ambition, and a new way of understanding the world.

From Lyell, Darwin learnt to see geological change as perpetual and gradual. Most previous thinkers, even if they discounted the biblical time-span, still tended to assume that the earth's continents had been shaped over a relatively short period by cataclysmic events like the Flood. Lyell argued that slow, everyday forces of wind, rain, ice and sun had sculpted the earth's lakes, plains, valleys and mountains over eons of time. His subtitle for *Principles* said it all: *An Attempt to Explain the Former Changes of the Earth's Surface, by Reference to Causes Now in Operation.*[51]

Charles Lyell gave geology nothing less than a new set of professional scientific principles. These freed the study from squabbles over biblical time-scales that had previously marred its reputation and turned it into a political minefield for many naturalists. Without disputing divine creation, which Lyell believed to have set the machinery of geological change in action, he insisted that modern geologists must use only observable natural forces when explaining how this change operated in the world. To do otherwise was unscientific.

It was the vast age of the earth, claimed Lyell, that had enabled small forces such as erosion to produce enormous changes. Among the most important forces were the slow movements of elevation and subsidence in the earth's crust. Repeated volcanic eruptions and earthquake fractures had produced successive rises and falls in the height of surrounding land.[52] What looked to be static and perpetual was actually mobile and malleable. As Darwin wrote excitedly in his diary, 'the solid earth, considered from our earliest childhood as the very type of solidity has oscillated like a thin crust beneath our feet'.[53]

At the same time, Lyell argued, this process of change – in essence, a type of ceaseless geological flux – revealed no particular direction or purpose. The supposed greater complexity and perfection of fossils over time, for example, was wholly illusory – the consequence of an imperfect geological record. Despite the fantastic claims of transmutationists like Lamarck, there was not enough evidence to draw firm conclusions as to whether animals were more complex and developed in the present than in the past. Lyell, a firm creationist, stressed too that continual small shifts in the earth's structure could never alter the fixed biological character of species. These would remain essentially immutable from their moment of creation.

On the other hand, alterations in physical environments could cause some species to become extinct, and others to exhibit slight temporary variations of structure or behaviour in response to their new conditions. Such variations would disappear over time as a result of interbreeding among wild populations. They would never produce permanent transmutations.[54]

Becoming a Lyellian geologist was not easy or instantaneous for Darwin; it required a new way of thinking. He came to revere *Principles* because 'it altered the whole tone of one's mind, and therefore . . . when seeing a thing never seen by Lyell, one yet saw it partially through his eyes'.[55] Mastering Lyell's way of seeing in order to discern how invisible laws worked throughout the physical world came only with practice.

Even so, Darwin had made his first attempt at Lyellian

interpretation in early January 1832, only a month after leaving England. Gazing across the harbour of Porto Praya on the island of St Jago, he suddenly understood how a distinctive horizontal white band running through the cliff-face might have come into being. He decided that the stripe must consist of pulverised shells and calciferous matter that had been elevated gradually from the sea bed by volcanic activity.

For Darwin, this was a life-changing moment. The thrill of realising that he had come up with an original and plausible scientific hypothesis for the first time moved him to make an unexpected decision. On 7 January, while munching biscuits and ripe tamarinds on the beach at nearby Quail Island, he suddenly decided to write a book on the geology of the countries of the Southern Hemisphere visited by the *Beagle*. 'I do not think the impression this day has made will ever leave me,' he wrote in his diary.[56] It was on this day that Philos became a scientist.

Thanks to Lyell, geology tended to dominate Darwin's interests for the remainder of the voyage. The boy who at Edinburgh University had sworn never to read another book of geology now found that geological hypotheses swirled in his brain like 'a hurricane'. But it was not until the *Beagle* sailed through the Magellan Strait into the Pacific, on 10 June 1834, that Darwin found the opportunity to test these new Lyellian ideas on a large scale.

At first viewing, the landscape of the Pacific west coast of South America looked far from promising. Reaching Chile's chief seaport of Valparaíso on 23 July, however, made him realise how much he was beginning to see through Lyell's eyes. In some respects, the port reminded him of picturesque Santa Cruz viewed from the *Beagle*'s deck two years earlier, but he no longer looked at the scene in the manner of a poet. Instead, in his mind's eye, he saw the whole of Chile 'as on a map', bounded by the Andes on one side and the Pacific on the other. His imagination seized on the Lyellian idea that 'all powerful time can grind down mountains . . . into gravel and mud'.[57]

Then, in January 1835, while the *Beagle* was sailing around the large offshore island of Chiloé, Darwin stood on the deck to see towering Mt Osorno on the mainland suddenly belch out jets of smoke and lava. A few weeks later, moored slightly further north in the small mainland harbour of Valdivia, he learnt that the earlier volcanic eruption had been followed by 'the most severe earthquake which the oldest inhabitants can remember'. The *Beagle* arrived at Valdivia in the immediate aftermath. Though Darwin was appalled at the spectacle of flattened houses and smashed buildings, he could hardly contain his excitement at being so close to a major geological event.

Sailing further up the coast to Concepción a few days later, he surveyed the damage created by the epicentre of the quake, which had fissured land for four hundred miles around. Locals reported that the aftershock had generated a 'great wave, which travelling from seaward burst over Talcuhano. In the middle of the bay it was seen as one unbroken swell of water: but on each side, meeting with resistance, it curled over, and tore up cottages and trees as it swept onward with overwhelming force.'[58] Darwin was seeing the after-effects of his first Pacific tsunami.

As it happened, it was FitzRoy who made the detailed geological observations at Concepción, and he later produced a paper that proved he shared Darwin's commitment to Lyellian principles. Darwin noted that his captain brought a mass of evidence to show that the primary earthquake had been accompanied by numerous smaller ones, each of which 'by lesser and even insensible steps' drove the surrounding land another step higher. Darwin calculated that there had been perhaps three hundred of these 'separate strokes', and he inferred that volcanos were associated with similar processes of surrounding uplift. He made the pithy observation that, 'in a volcano the lava is ejected, while in the formation of a mountain chain it is injected'.[59]

After a short stay in Concepción they sailed back to Valparaíso, where Darwin had time to make a more sustained inland expedition to the mountain chain of the Andes, to see if his Lyellian theorising stood the test of empirical scrutiny. Clambering up the western line of the Andes on 20 March, he discovered shells and other marine remains at a height of 13 210 feet. By now he had read the second volume of *Principles*, in which Lyell stressed the way that geological barriers influenced the distribution and character of flora and fauna (without of course altering the fundamental fixity of species). Descending on the eastern side of the Cordillera mountain range, Darwin realised he was in the middle of a perfect illustration of Lyell's thesis. The valleys on the eastern side of the Andes had developed markedly different vegetation from those on the western side, despite their relatively similar climates, soils and animal species.

His mind began to probe the implications of this, taking him another small step towards what would eventually become his theory of evolution. He hypothesised that the Andes had existed as a geographical barrier for such a length of time that 'whole races of animals must subsequently have perished from the face of the earth'.[60]

In a diary footnote, Darwin called this latest insight 'an illustration of the admirable laws first laid by Mr Lyell', but his luminous imagination had already begun to push his guru's ideas in fresh directions. Seeing fossilised trees embedded in the strata of the Andean mountaintops instantly unfolded a 'marvellous story' in his mind. Where he was standing, clusters of fir trees had once waved their branches on the shores of the Atlantic, now seven hundred miles away. Volcanic eruptions, accompanied by elevation and subsidence, had eventually produced the landscape around him, 'an utterly irreclaimable' desert, where 'even the lichen cannot adhere to the stony casts' of former trees. He told his sister Susan that he could see 'manifest proofs of excessive violence, the strata of the highest pinnacles are tossed about like the crusts of a broken pie'.[61]

With his mind sweeping forward over the entire western coastal terrain of South America, then back through the recesses of deep time, he proposed a major scientific hypothesis:

> I have certain proofs that this part of the continent of South America has been elevated near the coast, at least from four hundred to five hundred feet since the epoch of the existing shells and further inland the rise may possibly have been greater . . . At a remote geological era, it is probable that the Andes consisted of a chain of islands, which were covered by luxuriant forests.[62]

What he saw, plain as day, was not a Humboldtian poetic landscape of the present, but an invisible Lyellian landscape of the past.

Deep down, Darwin knew that he had at last found his scientific wings. Though he would not yet have used the term, he had become a biogeographer – one of the vital intellectual steps on his way to being an evolutionist. 'I cannot express the delight which I felt at such a famous winding up of all my geology in South America,' he wrote to his sisters.[63]

Soon after, he arranged to make one last South American land-based journey, up the Pacific coast for four hundred miles to meet up with FitzRoy and the *Beagle* at the small port of Copiapó. It was a measure of how far he had travelled psychologically since leaving England four years earlier that he was now beginning to feel 'tired of this eternal rambling' and glad to rejoin his ship and companions.[64]

# Islands on His Mind

As the voyage entered its final phase before turning back towards the Northern Hemisphere, Darwin found himself overwhelmed by nostalgia, an affliction that Captain Cook had associated with southern voyaging. The long chatty letters from his sisters that had entertained him with Jane Austen-like morsels of gossip about local Shropshire flirtations and marriages contained one piece of news that hurt him badly. Lovely Fanny Owen, the family friend with whom he'd played fantasy games in the woods, she as Housemaid and he as Postillion, had unexpectedly married.

Before leaving England, and with encouragement from her, he'd imagined Fanny as 'a nice little wife for the Parsonage'.[1] Now his dream of a quiet home in the English countryside was untenanted by a wife, and he longed to find someone to take Fanny's place.

Darwin could be pardoned for feeling tired. His body had been weakened by an illness at Valparaíso that some historians have thought to be Chagas' disease, a mysterious ailment transmitted by the bite of an insect. The rolling Pacific swell also upset his stomach and depressed his spirits. Even the meridian of the Antipodes on 19 December 1835 passed like a mirage. 'I looked forward to this airy barrier as a definite point in our voyage homewards; now I find it and all such resting places for the imagination are like shadows, which a man moving onward cannot catch,' he wrote in his diary. He could not help feeling vexed, after such a prolonged period of sailing, by 'the want of room, of seclusion, of rest – the jading feeling of constant hurry – the privation of small luxuries, the comforts of civilization, domestic society . . .' The Pacific Ocean loomed like 'a tedious waste, a desert of water'; its blue expanses were welcome only for giving him 'the time and inclination to measure the future stages of our long voyage of half the world, and wish most earnestly for its termination'.[2]

This ennui was to taint Darwin's appreciation of many of the places he visited in the remaining year of the voyage – just as it has dulled the evaluation of many historians who have followed in his tracks. Consciously or otherwise, Darwin compared these southern landscapes with a rosy ideal of home. Nostalgic blinkers immunised him to New Zealand's beautiful Bay of Islands. Oceans of waving ferns seemed 'rather desolate', the best vistas 'only occasionally pretty'. Gardens at the missionary village of Paihia were 'quite pleasing' only because of their 'English flowers . . . roses of several kinds, honeysuckle, jessamine, stocks and whole hedges of sweet briar'. On departing New Zealand on 30 December 1835, he grumbled that he was 'glad to leave'.[3]

Australia and South Africa fared no better. Arriving in New South Wales at the end of a drought-ravaged summer didn't help, and the young man who'd seen a temple-like beauty in the wastes of Patagonia could find little to appreciate in novel southern landscapes. Riding westward from Sydney, he complained of 'the arid sterility' of the trees, the monotony of eucalypt foliage, dried-up riverbeds and 'wearisome' sandstone plains. Van Diemen's Land was better only because it was greener and more English-looking. King George Sound, in present-day Western Australia, appeared 'uninviting'. On 14 March 1836, he left Australian shores 'without sorrow or regret', and a few months later he complained that Cape Town's lush hinterland was the most uninteresting country he'd seen.[4]

These caustic judgements are unfortunate because they have blinded many commentators to the intellectual significance of Darwin's last year of travel in the southern oceans. With the exception of his visit to the Galapagos Islands in October 1835, which historians invariably regard as canonical, the key achievements of Darwin's voyage are assumed to be over once the *Beagle* heads south for Tahiti.

This is a serious oversight. Some of his most important discoveries were to come among the islands and island continents of the Indian and Pacific oceans. Homesickness might have made him grumpy about the beauty of these landscapes, but it didn't affect his passion to discover new natural laws. Indeed, the Galapagos visit was only the preamble to an integrated sequence of explorations that opened up entirely new horizons of knowledge and speculation. Darwin quickly appreciated that islands were good to study because of their manageable size; their recent origin as landforms, habitats and societies; and their perplexing organic relationships with neighbouring landmasses.

Darwin's reading of the second volume of Lyell's *Principles* had given him a definite, if loose, agenda for these forthcoming investigations. The islands of the Indian and Pacific oceans, with their confined habitats, would simplify his mission to collect comprehensive samples of plants, birds, insects and animals, and to track their migrations. As in the Andes, he was to use the 'uniformitarian' insights and methods of a Lyellian geologist to investigate the physical origins of islands and archipelagos. This meant finding evidence of how slow, everyday and sometimes invisible actions had brought these fragments of land into being, including the forces of volcanic uplift, ocean-floor elevation and subsidence, and coral growth.

Lyell's book had also alerted Darwin to some of the ways in which recently formed landmasses could become populated by new networks of flora and fauna, able to adapt to local climate and soil. Darwin expected to find some slight and transient variations among these chance strays from mainland 'centres of creation'.

As well – and here Darwin's agenda was as much influenced by his experiences with Fuegian natives in Patagonia – he would investigate how these processes of migration, dispersal and adaptation influenced human populations and societies. In particular, he'd observe how new influxes of peoples, or 'invasive species', often from 'civilized' northern centres such as Britain, had influenced the original 'primitive' inhabitants of southern islands or continents and reshaped their natural environments.

Finally, the second volume of *Principles* had set Darwin thinking about how coral reefs were formed. Lyell, knowing that corals could survive only in relatively shallow water, had along with several other theorists speculated that their polypi must grow on top of underwater volcanos that were being gently elevated from the ocean floor. To Darwin's mind, though, this failed to explain the different shapes of the three prevalent types of coral reef: barrier reefs, which arose some distance from the shore, their outer sides growing out of deep water; fringing reefs, which grew around volcanic islands; and atoll reefs, which formed a ring or horseshoe shape, with a lagoon nestling inside and drifts of sand and dead coral coalescing into a new 'coral island'.[5]

Bizarrely, even though Darwin had yet to see a coral reef, he believed he'd already solved this puzzle. While viewing the vast land elevations on the west coast of South America, he had suddenly thought that there might be corresponding subsidence occurring elsewhere on the seabed. Could this be the explanation for the various types of coral reefs and islands that were known to stud the Pacific and Indian oceans? Had coral reefs grown upward on top of land that was subsiding slowly under the sea? Sailing among the islands, archipelagos and island continents of the South Seas offered Darwin a set of unique, living laboratories for testing this wildly deductive hypothesis and for honing his knowledge and skills as a theorist of geology, zoology and ethnography.

And though he did not know it at the time, the results of his investigations here would push him several steps closer to a theory he would one day call 'modification by descent through natural selection'. Today we know it simply as evolution.

By the time the *Beagle* arrived at the Galapagos Islands on 15 September 1835, after sailing some six hundred miles west of Ecuador, Darwin was absorbed in the accounts of several earlier visitors, including the naval explorer John Byron, who'd made a brief food stop there on HMS *Blonde* in 1825. Byron had reported that 'the birds and beasts do not get out of our way . . . and all this amidst volcanos which are burning around us . . .' It was 'like a new creation'. Darwin was equally excited to learn that the naturalists who had visited the Galapagos had all failed to amass major collections. It was for this reason that Sir William Hooker, the famous Glasgow botanist, had lobbied the Admiralty to include the Galapagos on the *Beagle*'s itinerary.[6]

Even though the landscapes were as melodramatic as Byron had suggested, the geology of the Galapagos proved more straightforward than Darwin hoped.[7] At their first stop, Chatham Island, the blackened coast reminded him of industrial Wolverhampton in the English Midlands. He and FitzRoy were also struck by the place's resemblance to hell, especially vivid in Darwin's mind from the copy of Milton's *Paradise Lost* he carried in his back pocket. The beach and hinterland were littered with broken lava heaps and scorched boulders. Scraggy plants struggled like weeds among the clinkers. Volcanic craters pitted the ground at every point – more than two thousand of them in Darwin's estimation. They ranged in size from mounds to mountains, some made up of solidified scoriae and lava, others sculpted out of finely stratified sandstone. Cone-shaped formations stood among them like sentinels: Darwin decided that these too were heaps of solidified volcanic mud.

When on 21–22 September he managed to scramble up the higher volcanic mountains on the northern part of the island, Darwin was disappointed to find that their craters were dormant. Rings of cinders marked where they'd once belched molten lava.[8] He'd longed to study an active volcano, but even Narborough, one of the most recently elevated islands to the west, had stopped erupting since Byron's visit. Now it gave off only a thin jet of steam. At least Charles Island, populated by an Ecuadorean penal colony, managed to produce one satisfying Lyellian moment. After climbing two thousand feet up its central mountain, Darwin worked out that the lava must originally have flowed from under the sea. Having been elevated by volcanic uplift, it had now eroded into fertile black soil.[9]

But despite all that he'd read, Darwin had not expected organic life on the Galapagos to be so strange. It seemed to come from another planet. He couldn't imagine what centre of creation could have produced such a diabolical array. Some birds and plants he recognised as having South American affinities, but what were penguins doing among the overgrown lizards in this tropical habitat? And why had the creator, whose beautiful designs Darwin's Cambridge textbook by Archbishop Paley had trumpeted, chosen to create a paradise for reptiles? And these were not just ordinary reptiles. There were tortoises weighing more than two hundred pounds that needed eight men to carry them, and 'disgusting clumsy lizards', or iguanas, that

Darwin called 'imps of darkness'. One species of iguana fed off the land, the other grazed on underwater seaweed – as Darwin proved when he dissected its stinking guts on the poop-cabin table.

In the interests of science he repeatedly threw a seagoing specimen into the water to find out whether its instinct was to return to land. Sure enough, the creature patiently clambered back on shore each time, thereby implying that it had adapted itself to finding aquatic food. Darwin could see that these Galapagos reptiles were fulfilling the role of mammals in other parts of the world, but why?

The exoticism of the plant and animal life galvanised him into a collecting frenzy. On James Island, where they stayed for a week from 8–17 October, he, Covington, assistant surgeon Benjamin Bynoe and others shot, cut down, dug up, gathered, fished and trapped specimens from many branches of natural history. Darwin even camped on the beach so as to start work before dawn. He forced himself to gather specimens from the rugged highland interiors of the island, where no naturalist had previously reached.

Captain FitzRoy did some collecting too, amassing less than Darwin but, as it turned out, labelling his specimens more accurately. Darwin collected blindly, more interested in obtaining a large sample than in recording where he'd found particular items. He thought, for example, that several hundred plant specimens he'd gathered were duplicates, not realising they were distinct species. Neither did he know that the numerous little birds he shot, today reverently called 'Darwin's finches', were finches at all, let alone more than twenty different species of the genus. He did notice some differences in their beaks, but he treated these as insignificant deviations within a single large and widely dispersed group. This oversight was to make the whole sample useless for scientific citation.[10]

Two anomalous pieces of information did get an entry in his notebooks, though Darwin was too busy to think about their implication. Mr Lawson, the governor of Charles Island, told him that locals could tell from the shape of the shells which particular island the tortoises came from. Darwin himself also thought it 'singular' that mockingbirds on Chatham, Charles and James islands were each 'consonant with its own island'.[11] But at the time, these observations seemed to him no more than mildly interesting; they were, after all, perfectly compatible with a creationist belief that small and temporary variations could arise from local adaptations.

All in all, when the *Beagle* set sail for Tahiti on 18 October 1835, Darwin's satisfaction came more from his large specimen collection than from any new natural-history insights.

For the next twenty-five days the *Beagle* surfed 'the boundless ocean', until on 15 November they came in sight of Tahiti, 'an island', Darwin noted, 'which must forever remain as classical to the Voyager in the South Sea'.[12] Oddly, the jaded sentiments he'd expressed to his sisters about this last leg of the voyage had not been replicated in exchanges with naturalist friends. He told his naturalist cousin Henry Fox that he was looking forward to visiting Galapagos, Tahiti, Sydney and Cape Town 'with more interest than the whole voyage'. And his geologist friend Robert Alison, then visiting Valparaíso, wrote of his own eagerness to hear Darwin's report on Pacific islands: 'it will be curious if you find a sinking of the land there & a rising here'.[13]

It's true that Darwin did not expect to find many new specimens. His time on land would be fleeting, and anyway most of the places they intended to visit had been well worked over by the likes of Robert Brown, botanist on Matthew Flinders' voyage; the two Cunningham brothers, who'd worked for many years in New Zealand and Australia; and the distinguished entomologist William Sharp Macleay, who'd moved permanently to Sydney. No matter,

Darwin told his sister Caroline, he intended instead to write 'much about the missionaries'.[14]

This shift in his intentions was opportunist, but in no way trivial. Having seen FitzRoy's Patagonian mission collapse, Darwin was eager to explore the impact of missionaries elsewhere. He would embark on a series of comparative investigations into how the social impact of agents of English 'civilization' – settlers and missionaries – had influenced the original 'primitive' inhabitants and the environments of the islands they had colonised. Having shifted from rocks to animals, Darwin was now moving to people.

As usual, he looked for hypotheses to test by reading in advance about each prospective destination. For Tahiti, his authorities were F. W. Beechey's *Narrative of a Voyage to the Pacific and Beering's Strait* (1831), William Ellis's *Polynesian Researches* (1829), and Otto von Kotzebue's *Voyage of Discovery in the South Sea* (1821). Beechey, a naval captain, was neutral about the missionaries' impact; Ellis, a missionary himself, praised his colleagues; and Kotzebue, a Russian sea captain with radical views, attacked them for crushing the Tahitians' vitality and freedom, leaving the people gloomy and repressed.[15]

During the fortnight that the *Beagle* spent in Tahiti, Darwin found the locals as delightful as voyager eulogies had claimed. The men, in particular, seemed gentle, intelligent and civilised. Their athletic brown bodies put the white men to shame, making them seem bleached and insipid. The universal Tahitian adoption of Western clothes, so disparaged by some observers, seemed perfectly dignified, and their tattooed bodies were beautifully 'ornamented like the trunk of a tree by a delicate creeper'.[16]

He saw no signs of imposed inhibitions: a group of children sang for the sailors on the beach with charming spontaneity; a guide who'd led him up the volcanic peaks prayed earnestly before sleeping. Thanks to the missionaries, the vices of kava drinking, alcohol, infanticide, warfare and cannibalism had disappeared. The three missionaries he met struck Darwin as good men, more practical than pious. He was also relieved to see no obvious signs that the people's traditional food sources had been damaged: lush tropical fruits and vegetables still grew with wild abundance and were cooked in the delicious local manner reported by every voyager from the time of Cook. When the *Beagle* set sail for New Zealand on 26 November, Darwin could only add his quantum of praise 'for the island to which every traveller has offered up his tribute of admiration'.[17]

The *Beagle* spent only a few weeks at the next destination, New Zealand's North Island, which it reached on 21 December 1835. From the Bay of Islands, Darwin visited the nearby town of Kororareka and the missionary settlements of Paihia and Waimate. Here again, to his intense later regret, he concentrated on investigating the missionary impact rather than tracing the island's patterns of flora and fauna. He would one day have to expend much labour rectifying this omission.

FitzRoy, too, encouraged him to focus on missionaries. The captain was furious at the opinions of former *Beagle* artist Augustus Earle, a man whom Darwin had liked in spite of his bohemian ways. He and Earle had shared premises in Brazil before illness forced the artist to leave the voyage. Darwin also shared Earle's strongly emancipist views on Brazilian slavery. Some years earlier, however, Earle had published the vehemently anti-missionary *Narrative*, about the nine months he spent in New Zealand in 1827, a book that Darwin and FitzRoy read on the *Beagle*. Both now agreed that the vice-ridden painter had been a treacherous witness against his generous missionary hosts.[18]

Darwin's ethnographic observations also bore the marks of local missionary influence. In every respect, he claimed, the Maori compared adversely to the Tahitian: 'One is a savage, the other is a civilized man.' The eyes of Maori warriors were cunning and ferocious, their figures bulky, their bodies and houses filthy. He could see no principle of government among and between the various Maori tribes other than the use of violence: 'the love of War was the one and lasting spring of every action'.[19] Informants told him that polygamy was common and cannibalism not yet extirpated. Overall, he placed Maoris on the scale of savagery only just above the Fuegians.

Darwin did concede, however, that missionaries had often been forced to depend on Maori people for protection against the worse violence of settlers, many of them ex-convicts from Australia, who were 'the very refuse of society, completely addicted to drunkenness and vice'. He noted too that settlers had razed kauri forests for pasture and introduced devastating pests such as rats, wild leek, common dock, and other weeds. Only the missionary establishments, with their English flora and manners, softened what for him was 'not a pleasant place', and from which he gladly departed on 30 December.[20]

Just under two weeks later, on 12 January 1836, the *Beagle* entered Port Jackson and anchored at Sydney Cove. Darwin's initial response was enthusiastic; he later admitted that his expectations of New South Wales had been utopian. He'd even nursed a fantasy of one day migrating there in search of gold. Even so, he was staggered to see a glittering city of 23 000 people that exceeded London and Birmingham in the speed of its growth. There were grand sandstone buildings, well-furnished shops, and carriage-filled streets. More civilisation seemed to have been achieved here in 'a score of years' than in South America over the same number of centuries: it made him proud to be an Englishman.

Yet this early impression waned, especially as Darwin came to share the conservative opinions of Sydney's free-settler elite. On 26 January, he visited the influential family of *Beagle* midshipman Philip Gidley King at their Dunheved homestead, near today's

Penrith, thirty miles west of Sydney. Here Darwin viewed Captain Philip Parker King's estate and discussed naturalism and the condition of the colony. The next day, he took lunch with the Kings' relatives by marriage, the Hannibal Macarthurs, whose opulent Palladian property overlooked the Parramatta River. Both families were intensely hostile to 'emancipists', as ex-convicts were called.

Darwin's chief interest in the colony had been to assess the character and impact of a unique, convict-based British society, an invasive species like no other. Everything he learned, however, outraged his gentry sensibilities. Ex-convicts had become some of Sydney's wealthiest men; they rode in gilt carriages, built extravagant houses, and warped the character of the whole society. Rich or poor, they were obsessed with money: bookshops, fine music, and other aspects of civilisation had no appeal. Darwin anticipated that the gross and sensual manners of convict servants would rub off on future generations and lead to an overall moral decline.

At the garrison of Bathurst, a few days' ride from Sydney, disgruntled soldiers taught him a harsh social typology. A 'squatter' was a freed convict who'd turned farmer, built a bark hut, and become rich trading in stolen goods and illegal spirits; a 'crawler' was an assigned convict who'd run away to live by petty theft; and a 'bushranger' was 'an open villain, who subsists by highway robbery'.[21] It seemed an unpromising basis for a future society.

Thankfully, the smaller town of Hobart, capital of the southern island of Van Diemen's Land where the *Beagle* stayed in early February, boasted fewer rich convicts and a more English social tone. A dinner with the naturalist surveyor-general George Frankland proved one of the most enjoyable evenings Darwin had experienced since leaving England. All in all, though, he doubted whether Australia's harsh climate and flawed social structure would enable it to become a second America. The best Australia could hope for would be to 'reign a great princess in the South'.[22]

Yet Darwin thought unusually well of the Aboriginal inhabitants of New South Wales. While on an excursion to the Blue Mountains on 16 January, he bumped into a small party of young men. With their spear-throwing skills, their 'good humoured and pleasant' countenances and their acute observations, 'they appeared far from such utterly degraded beings as usually represented'.[23] However, anticipating his later views on Malthusian-style struggles, Darwin feared for their survival because of intertribal warfare, a wandering way of life, high infant mortality, and exposure to European spirits and disease. On top of this, the extinction of wild animals by European guns, dogs and land policies was eroding traditional sources of sustenance. The consequences looked ominous: 'when the difficulty in procuring food is increased, of course the population must be repressed in a manner almost instantaneous as compared to what takes place in civilized life'.[24]

The situation in Van Diemen's Land was, if anything, worse: population decline had been hastened by the forced isolation of Aborigines on a promontory, where they lived 'in reality as prisoners'. If 'this cruel step' had become unavoidable because of settler–native conflict, 'without doubt the misconduct of the Whites first led to the Necessity'. Though Darwin did not yet think to transfer to the zoological world these same ideas of population control through struggle, he was clearly laying down the elements of his future theory of the survival of the fittest.[25]

Only at King George Sound on the far west coast of Australia, where the *Beagle* called for eight days from 6 March, did he sense that Aborigines were defying the trend towards extinction. A fragile white settlement without convict labour, King George Sound relied on the goodwill and work of the local Aboriginal people. Darwin thought them good-natured, hardworking and physically robust. As a result he decided to instigate an experiment in black–white 'conciliation' by providing rice and sugar for a visiting group of

Cockatoo men to stage a corroboree with their local counterparts. That evening, the two groups danced in front of a flickering fire, naked apart from painted white spots and lines on their bodies, all stamping their feet and chanting in unison. FitzRoy found the sight 'fiendish'.[26]

Darwin, by contrast, loved the dancers' zest and their clever mimicry of an emu hunt. He'd seen such festivities in Tierra del Fuego, but 'never where the natives were in such high spirits and so perfectly at their ease'.[27] On a spectrum of 'barbarism', he placed the likeable Aborigines above both the Maori and the Fuegians, but below the Tahitians.

On the whole, Australia's geology disappointed him. He misinterpreted the origins of the Blue Mountains, thinking their vast, bay-like ravines and valleys to have been carved by an ancient sea. He surmised that rainfall erosion would have taken far too long to generate such spectacular effects. Lyellian he might be, but Darwin was still badly underestimating the age of the continent.[28] At Bald Head, near King George Sound, he did, however, see evidence of a major coastal elevation, which suggested the likelihood of compensating subsidence of the ocean floor elsewhere in the Pacific.

As in the Galapagos, it was the strangeness of Australia's animals that nagged most at Darwin's mind. On 19 January, these reflections led him to speculate on a dangerous new subject, the theologically sensitive issue of the origin of species. It would be one of the great understatements to call this a portentous moment. At no other time on the *Beagle* voyage did Darwin raise the issue, and afterwards he buried it for a further twenty years.

Early that morning, the superintendent of a farm at Wallerawang, west of the Blue Mountains, took Darwin out on a hunt. After a poor showing, they managed to shoot a specimen of Australia's most extraordinary creature, the platypus. While swimming in the creek, it had looked and behaved, Darwin thought, exactly like a commonplace English water rat. Up close, however, its soft, bird-like beak was even stranger than on stuffed specimens he'd seen. The oddness of the creature was well known to English science, thanks to the work of London comparative anatomist Richard Owen.[29] But why, Darwin pondered, was this animal so different from the English water rat, whose role in the economy of nature it seemed to replicate?

A few hours earlier he'd been resting on a sunny bank, waiting for his host, when he began ruminating on 'the strange character of the Animals of this country as compared to the rest of the world'. At the time he'd been thinking particularly about parrots – birds so strikingly unlike any in England. On the other hand he'd also seen Australian crows and magpies that closely resembled their English counterparts. What did all this mean?

His diary takes up the story:

> An unbeliever in everything beyond his own reason might exclaim, "Surely two distinct Creators must have been [at] work; their object however has been the same & certainly the end in each case is complete". While thus thinking, I observed the conical pitfall of a Lion-Ant: – A fly fell and immediately disappeared; then came a large but unwary Ant; its struggles to escape being very violent, the little jets of sand described by Kirby (Vol I. p.425) were promptly directed against him. – his fate however was better than that of the poor fly's:– Without a doubt this predacious Larva belongs to the same genus, but to a different species from the European one. – Now what would the Disbeliever say to this? Would any two workmen ever hit on so beautiful, so simple & yet so artificial a contrivance? It cannot be thought so. – The one hand has surely worked throughout the universe. A Geologist perhaps would suggest that the periods of Creation have been distinct & remote the one from the other; that the Creator rested in his labour.[30]

This is an often-quoted passage, though difficult to interpret because of its ambiguous phrasing, complicated by Darwin's concern that his Christian family and friends might one day read it.[31] He imagines a dialogue between two opposing theories, one which hypothesises separate creations in separate places, the other which assumes that the little Australian ant lion, almost identical to its British counterpart in form, function and behaviour, must surely have come from the same mind, suggesting that all species proceeded from a single creation in a single place. On the face of it, both these positions could be held by creationists, many of whom were divided on the issue of single versus multiple creations, and on single versus multiple centres of creation. Darwin's favoured position seems to be that of a Lyellian geologist – there had been more than one separate creation over time, interspersed by periods of natural geological activity.

Though Darwin nowhere advances a purely secular interpretation, his use of the phrase 'unbeliever in anything except his own reason' certainly implies it. What we can say with confidence is that he was beginning to question the variety of positions adopted by creationist naturalists, none of which satisfactorily answered the problem posed by Australian species. Darwin was in no way foreshadowing evolution, but he was gently deconstructing the logic of creation theory.

Darwin's Wallerawang deliberations also reflected the growth of his insight into the function and behaviour of species within the earth's overall ecology. Australian and British ant lions might be different, but Darwin believed that they shared the same 'office' or 'place' within the interdependent 'economy of nature'.[32] However, where no closely resembling species existed to fill such 'offices', they could evidently be taken by very different substitutes, such as the reptiles of the Galapagos or the marsupials and parrots of Australia. But why and how such species were driven to fill these vacant offices remained unclear.

Ecological puzzles were thus edging Darwin towards an evolutionary hypothesis, though he did not then realise it. Neither did he realise that the next and final leg of the *Beagle*'s voyage, where he shifted his attention from people, parrots and platypuses to tiny coral polypi, would take him further and faster down that dangerous track.

On 1 April 1836, the *Beagle* reached the isolated archipelago of the Cocos (Keeling) Islands in the East Indian Ocean, halfway between Australia and Africa. Now administered by Australia, the archipelago was then ruled by a commercial adventurer in coconut products, Captain John Clunies Ross. Darwin was unusually excited by the visit, which he probably persuaded FitzRoy to make especially for him. The prospect of testing the theory that had been buzzing around in his mind for six months swept aside any thoughts of homesickness.[33]

He had first glimpsed a coral reef from the masthead of the *Beagle* in November 1835, as they sailed past the 'Dangerous Archipelago' (Tuamoto). At nearby Tahiti, he'd been able to make a more leisurely examination of the local barrier reefs from a series of mountaintops – especially Eimeo Island, with its ring of coral and its glassy lagoon. When the *Beagle* turned towards New Zealand, this rudimentary fieldwork had emboldened him to draft, in December 1835, a rough explanation of how coral reefs might be formed.

He speculated that a series of subsidences had taken place in various parts of the Pacific, 'of which no one exceeded in depth, the number of feet which saxicolous polypi will flourish. The intervals were sufficiently long to allow their growth.'[34] In short, the speed of coral growth had kept pace with the gradual dropping of the ocean floor, immersing the live coral always in relatively shallow water where they were able to grow. New coral growth thus rested on layered ramparts of dead coral that had been unable to survive in the inky deep water.

Darwin knew he had to put these deductions to an empirical test. Spending eleven days among the various atolls of the Cocos Islands offered him a perfect opportunity, one made all the more pleasurable by the tropical beauty, which soon reinvigorated his poetic feelings. He found the spectacle of the atolls, viewed from the ship's northern anchorage, breathtaking. The white sand islands rose barely thirty feet out of a vast, rolling southern ocean, creating a giddying effect for those on shore. Inside the reef:

> the . . . clear and still water of the lagoon, resting in its greater part on white sand, is when illuminated by a vertical sun of a most vivid green. This brilliant expanse, which is several miles wide, is on all sides divided either from the dark heaving water of the ocean by a line of breakers, or from the blue vault of Heaven by the strip of land, crowned at an equal height by the tops of the Cocoa nut trees.[35]

Over the next ten days, he several times reawakened his old 'tropical fervour', particularly when drinking green coconut juice in bowers created by entangled palm fronds. On Horsburgh Island, some partial burning of the vegetation made his geological observations easier, and seemed to confirm the gradual subsidence he'd expected. He waded with Covington in the translucent lagoon shallows to collect the delicate corals of the inner reef, while coloured fish flicked in and out of the hollows and gullies. He vaulted with a pole to reach the outer reef's menacing margins, where waves crashed too heavily for him to wade. Here, as he'd hoped, he found that exposed corals had died from the effects of the sun, but those bathed by the sea were living. Also as predicted, the pieces he dragged on hooked lines from deep underwater were dead from lack of light.

Later, when they were around a mile from the shore, FitzRoy took soundings with a 7200-foot line and found no bottom. This suggested that the reef formed a vast submarine mountain with sides steeper than a volcanic cone.[36] It followed, as Darwin explained in his diary, that a fringing and barrier reef could actually become an atoll: 'In time the central land would sink beneath the level of the sea and disappear, but the coral would have completed its circular wall. Should we not then have a Lagoon Island? Under this view, we must look at a Lagoon Isd as a monument raised by myriads of tiny architects, to mark the spot where a former land lies buried in the depths of the ocean.'[37]

When the *Beagle* sailed out of the lagoon on 12 April 1836, Darwin felt fulfilled: 'I am glad we have visited these Islands; such formations surely rank high amongst the wonderful objects of the world. It is not a wonder which at first strikes the eye of the body, but rather after reflection, the eye of reason.'[38]

For him the eye of reason had done more than confirm an intriguing geological hypothesis: his time on the Cocos Islands had also produced a unique body of evidence about the ecological functioning of atolls that was to prove hardly less significant for his future theory of evolution than the Galapagos visit.

Today we would call what Darwin discovered at Cocos (Keeling) an ecosystem. He saw how organisms had influenced their environment and vice versa. Tiny coral polypi had brought into being landforms on which new organic networks developed. This time, moreover, he described exactly where each specimen of plant, fish, mollusc, crab, spider or ant was found. This time he could test whether isolation on separate islands had produced unique variations. Having made what he believed to be a complete collection of Cocos biota, he was struck by the overall paucity of species – something apparently common to all islands. This suggested that the dispersal of a few common species from, say, Australia or Sumatra, rather than a process of divine creation, had influenced the types of plants and creatures that found their way onto these remote places.

He also saw clear evidence that these refugee organisms adapted their behaviour to suit new environmental conditions. Coral polypi had curtailed their growth and altered their shapes according to the strength of the waves; giant robber crabs had developed an ability to line their burrows with coconut husks, and to crack open the nuts with their fierce claws.

Above all, Darwin began to use metaphors of struggle to describe the perpetual battle between sea and coral: 'the ocean throwing its water over the broad reef appears an invincible enemy, yet we see it resisted and even conquered by means which would have been judged most weak and inefficient'.[39] He had traced the process whereby a tiny organism, by purely natural means, managed to bring about a 'new creation' in the form of coral islands. On these sandy 'refuges for the destitute' flourised plants and creatures that had arrived by chance. What would 'an unbeliever in anything but his own reason' say to this?

Confident as he was about the broad lines of his coral reef theory, there remained one type of reef, the 'fringing' or 'shore reef', which Darwin had not seen. To his delight, the *Beagle*'s final destination supplied the deficit. Mauritius, situated in the East Indian Ocean around twelve hundred miles from Africa, was encircled by fringing reefs. Between 29 April, when they anchored at St Louis harbour, and 9 May, when they departed for Cape Town, Darwin made observations on the northern side of the island. Somewhat to his surprise, these produced a slight modification of his theory. Because these fringing reefs encircled a volcanic island, they appeared to have been subjected to occasional periods of elevation as well as subsidence.

Even so, nothing contradicted his general explanation that an integrated set of geological processes linked all three types. The paper on coral reefs which he read before the London Geological Society a year later summarised this sequential process: 'on this view, the three classes of reefs ought to graduate into each other . . . fringing reefs are converted into barrier reefs; and barrier reefs, when encircling islands, are thus converted into atolls, the instance the last pinnacle of land sinks beneath the surface of the ocean'.[40] Lyell, on hearing this thesis, 'danced about and threw himself into the wildest contortions, as was his manner when excessively pleased'.[41]

With the benefit of hindsight, we can see that this rather arcane study of the origins of coral reefs was more important than even Darwin appreciated. Modern scholar Howard Gruber has pointed out how much, in a formal sense, Darwin's coral-reef hypothesis anticipated his later theory of evolution by natural selection. Two years before he realised the zoological significance of Thomas Malthus's ideas on population, Darwin had already proposed that coral numbers were limited by an intense struggle with natural forces and other organisms, such as the sun, sea, water temperature and coral-eating fish. He was arguing, too, that physical geology, together with population growth and environmental struggle, could explain major facts of long-term geographical distribution.[42] In the words of another modern authority, this was a process 'of sequential and irreversible change of form through time'.[43] We might call this evolution.

Mauritius was the last southern island that Darwin investigated. After a short call at Cape Town in June 1836, where he dined with astronomer Sir John Herschel, the *Beagle* headed for the familiar seas of the Northern Hemisphere.

The work of the voyage, however, was not yet over. It was now that Darwin began to take intellectual stock. He reviewed his notes, cross-referenced key points, and made comparisons between the islands and continents he'd visited. As he did so, new connections fired in his mind. He thought about the dispersal of species to remote islands, the effects of isolation, the pressures of environmental change, and the possibilities of varieties becoming species. It was only now that he glimpsed some significance in the fact that Galapagos tortoises and mockingbirds appeared to be different on each island. Could they have descended from a common ancestor and then diverged? 'If there is the slightest foundation in these remarks,' he scribbled blandly in his shipboard notes, 'the Zoology of Archipelagos will be well worth examining; for such facts would undermine the stability of species.'[44]

Something else strange happened, now that Darwin's longed-for return was a reality. Nostalgia, that most perverse of sentiments, started to play fresh tricks. As they drew nearer to England, he realised he might never again see the islands of the southern oceans. At Mauritius on 5 May 1836, after riding on an elephant down a road bordered by mimosa and mangoes, he had found himself writing wistfully, 'how pleasant it would be to pass one's life in such quiet abodes'.[45]

Six months later, on 2 October 1836, the *Beagle* anchored at Falmouth, after five long years away. 'To my surprise and shame,' wrote Charles Darwin, 'I confess that the first sight of the shores of England inspired me with no warmer feelings, than if it had been a miserable Portuguese settlement.'[46]

# PART TWO

# Joseph Hooker and the Ross Expedition, 1839–43

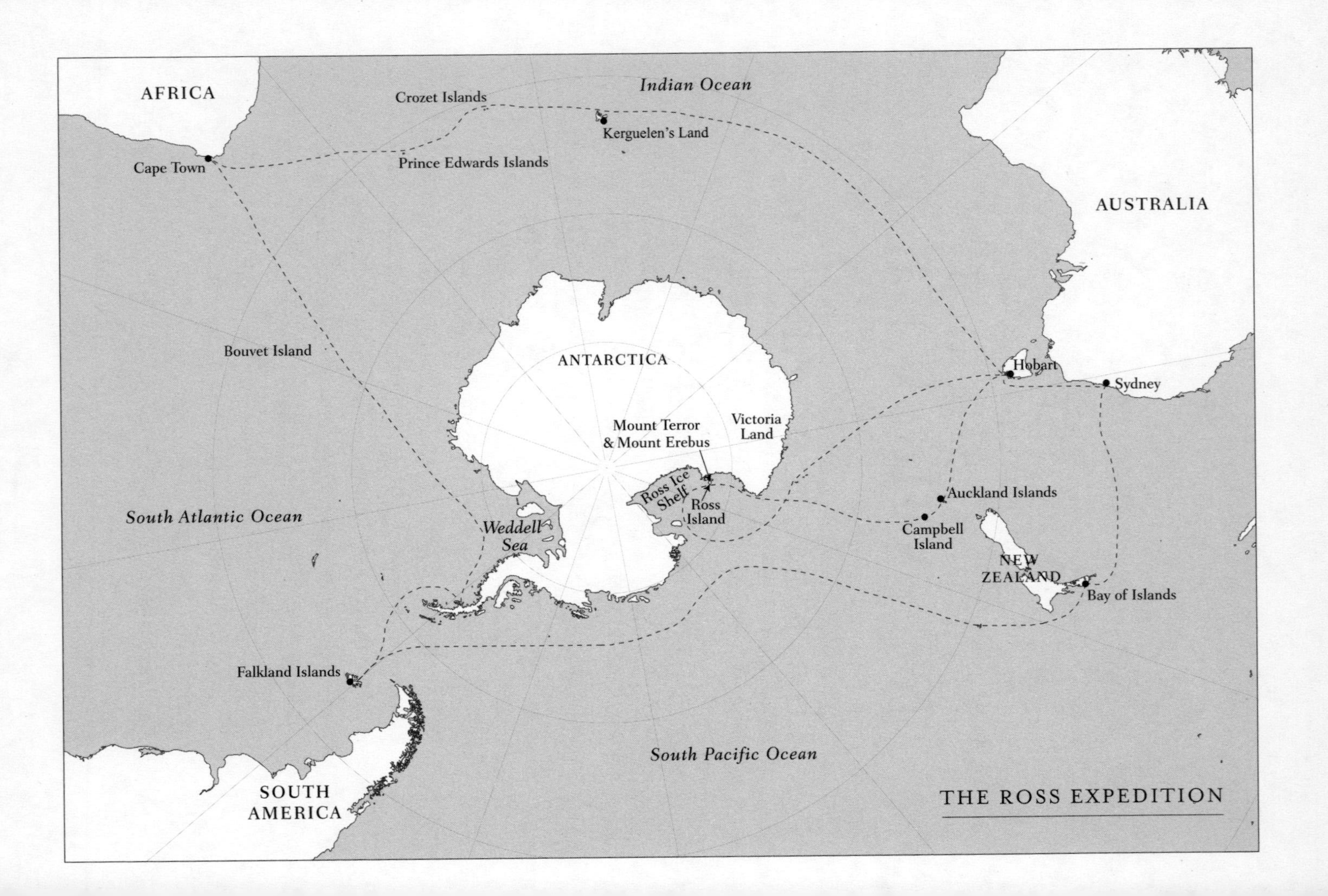
THE ROSS EXPEDITION
AFRICA
Crozet Islands
Indian Ocean
Kerguelen's Land
Cape Town
Prince Edwards Islands
AUSTRALIA
Bouvet Island
ANTARCTICA
Hobart
Sydney
Victoria Land
Mount Terror & Mount Erebus
Ross Ice Shelf
Ross Island
Auckland Islands
Campbell Island
South Atlantic Ocean
Weddell Sea
NEW ZEALAND
Bay of Islands
Falkland Islands
South Pacific Ocean
SOUTH AMERICA

# The Puppet of Natural Selection

In April 1839, Joseph Dalton Hooker, a 22-year-old Glasgow medical student, had an interview at Chatham docks with his potential employer, Captain James Clark Ross. Hooker had every reason to be deferential. He was a thin, short-sighted, pointy-faced young man with delicate lungs, and he was addressing Britain's most celebrated and experienced polar explorer.

Ross, said by some to be the handsomest man in the British navy, was tall and strongly built with an aquiline nose and a cockatoo crest of thick dark hair. He had joined the navy at the age of eleven, risen swiftly to the rank of captain, and served on seven expeditions to the Arctic regions, three of them in search of the fabled Northwest Passage to Asia. During one of these expeditions, when the steamship commanded by his uncle Sir John Ross was icebound in what is now the Canadian Arctic, James Ross set off behind a sled pulled by six Greenland huskies in search of the north magnetic pole. Using a single magnetic compass, he located the pole on a deserted limestone beach on 1 June 1831, at latitude 70° 5 17"N and longitude 96° 46' 45"W.[1] He well knew that the precise location of the meandering pole could not be fixed for more than a short time, so his silken British flag and cairn that marked the spot were not strictly accurate. Nevertheless he became a national hero, celebrated as 'Sir James Clark Ross, the first whose sole/Stood on the North Magnetic Pole'.[2]

Feared for his hauteur and hot temper, James Ross had a marked sense of rank, and Joseph Hooker, who had recently been appointed assistant surgeon on Ross's ship HMS *Erebus*, was of a lowly rank. Ross had promised this tyro position to Joseph several years earlier as a special favour to the boy's father, William Hooker. But far from being grateful, the boy was now insisting on being given a superior appointment, nothing less than the official naturalist of the two-ship expedition, which was to journey to the southern oceans and eventually into the Antarctic Circle.

Joseph Hooker had just heard that Ross had appointed Dr Robert McCormick, former Arctic veteran and ex-surgeon on the *Beagle*, to be surgeon and zoologist for the expedition's second ship, HMS *Terror*. Hooker was certain the man had been foisted on Ross, but being ignorant of Admiralty conventions, he failed to understand that the position of 'official' naturalist was almost always given to the senior medical officer, and of course there was no way that a raw, barely graduated young MD like himself would be awarded the post.

Joseph did know enough of shipboard protocol to realise that without official Admiralty recognition he would lose crucial naturalist privileges. McCormick would be entitled to go ashore, even at temporary landings, and, once there, would also have priority in collecting flora and fauna. Worse still, he would have the sole right, apart from the captain, to produce an Admiralty-supported publication of scientific observations. Hooker, as McCormick's junior, would have no automatic entitlement to shore leave and would have to hand over all his own notes and collections to the surgeon, who could use them as his own. Joseph's chances of achieving scientific fame were now fatally compromised. McCormick, who'd been so angered by Charles Darwin's favoured treatment on the *Beagle* seven years earlier that he'd resigned, would be likely to guard his rights jealously.[3]

In angling for a higher position, the young landlubber Hooker was being more provocative than he realised, especially when he hinted that he might make use of his father's connections and go over Ross's head to the Naval and Admiralty boards.

In response, Ross showed exemplary restraint, though he must have been nonplussed to have his command challenged by a glorified schoolboy. He explained that, were he to include a government-appointed naturalist on the voyage, it would have to be someone like 'Mr Darwin', who was 'perfectly well acquainted with every branch of Natural History' and 'well known in the world before hand'. Young Joseph Hooker, the captain implied, was emphatically not such a person. Unbowed, Joseph interjected hotly that Ross was wrong about Darwin: 'what was Mr D. before he went out? He, I dare-say knew his subject better than I now do, but did the world know him? The voyage with FitzRoy was the making of him (as I hoped this expedition would me).'[4]

At this point, Ross must have been tempted to terminate the argument by stating that a captain's word was final. Had Hooker been formally on the muster, Ross could technically have clapped him in irons. The captain could also have pointed out to this mutinous young 22-year-old that the expedition had no need of any naturalists at all: the Admiralty had appointed Ross himself both chief scientist and commander. This was hardly surprising, for he was far more qualified than either McCormick or Hooker. He'd served as naturalist on two earlier polar expeditions, accrued a substantial scientific reputation for his zoological, biological and astronomical expertise, was a Fellow of the Royal Society, and had just been elected a Fellow of the London Linnean Society. He had every intention, moreover, of taking personal charge of the dissection and microscope analyses of any marine life of the southern oceans collected on the voyage.

Ross also had far more pressing things to think about than the frustrations of an adolescent surgeon's mate. Natural history, much as he admired it personally, was not high on the Admiralty's priority list. Officially, the main object of this voyage was to investigate a more practical dimension of the science of navigation: what Ross himself had once described as the 'great and dark powers' of magnetism.[5]

Few were better qualified to understand those powers. As well as being the first man ever to see the magnetic needle dip vertically towards the North Pole, Ross had recently completed – in conjunction with the doyen of practical magnetic observation, Colonel Edward Sabine, FRS – an extensive survey of magnetic 'variations', or deviations from true north, around the British coastline. Both men regarded themselves as battle-hardened knights in what had come to be known as the 'magnetic crusade', a Europe-wide scientific mission to map magnetic behaviour across the globe.

Launched for navigational purposes, the magnetic crusade intersected with a more philosophical quest driven by curiosity. The applied and purely theoretical fascinations of magnetism attracted some of the finest scientific minds in the Western world. Scientists in Germany, France, Russia, Scandinavia and Britain were attempting to crack magnetism's three contradictory mysteries.

First and foremost was the fact that all compasses deviated a few degrees from true north as a result of the earth's rotational axis. If these precise deviations could be mapped all over the globe, they might be used as aids for calculating a ship's longitude at sea. Second, magnetism also possessed the capacity to draw a horizontal needle downwards towards the poles, a power known as vertical dip;

yet this too behaved mysteriously, in that the position of the north and south poles appeared to wander, so that the geographical and magnetic poles rarely coincided. The wandering nature of vertical dip had led some theorists to speculate that there might be up to four magnetic poles. Third, magnetic readings taken in the past showed that magnetic force changed in intensity over time. Regular readings were thus desirable in order for these changes to be taken into consideration.

The key to solving these mysteries seemed to lie in a global series of coordinated readings over short periods. To this end, influential scientists, headed by Alexander von Humboldt in Prussia and François Arago in France, had begun in the 1830s to urge the British navy to use its peerless resources to set up observatories in the untested magnetic regions of the Southern Hemisphere. British scientists, headed by John Herschel, himself newly returned from mapping the skies of the Southern Hemisphere, eventually persuaded the reluctant Whig government of Lord Melbourne to agree.

Herschel had explained in a letter to the Governor of South Australia why there were such pressing practical reasons, even apart from navigation, for amassing accurate magnetic readings. Given that surveyors often worked by compass alone, usually making no allowance or too much allowance for local variations in the earth's magnetic field: 'it is hardly possible to over estimate the amount of confusion and litigation which will be caused to the next generation when land becomes more valuable'. Colonial boundaries and settlements could degenerate into chaos.[6]

Everyone agreed that the expedition to set up the observatories should be led by the country's most eminent naval magnetician, Captain James Clark Ross. He was instructed to set up permanent stations at St Helena, the Cape of Good Hope and Van Diemen's Land, and to conduct sea-based observations throughout the southern oceans, especially among the high latitudes of Antarctica.

As it happened, Ross was well disposed to accede: he desperately needed the money that would come from such a prolonged period of employment. His would-be parents-in-law, the Coulmans of Whitgift Hall in Yorkshire, were sceptical about his recent engagement to their beautiful seventeen-year-old daughter, Anne. Ross, passionately in love, rightly suspected it was his poverty more than the disparity of ages that was troubling Anne's snobbish parents.

The Coulmans would not have been comforted to know of Ross's other, less advertised reasons for wanting to lead an expedition into the Antarctic Circle. Since the end of the Napoleonic wars, the French, Russians and Americans had begun to displace their fighting rivalries into geographical discoveries. Behind these, trade and imperial agendas also loomed. Indefatigable Sir John Barrow at the Admiralty Board longed for Britain to eclipse all these impertinent rivals, and he fancied James Ross was the man to do it. Within the Antarctic Circle lay great prizes for the intrepid or lucky explorer.

One was to journey further south into the icecap than had Captain Cook during the 1770s and several British whaling captains during the 1820s. Another, greater lure was to discover, map and claim the great southern landmass that was still thought to exist somewhere in those Antarctic climes – Barrow, one of the wildest of Britain's geographical projectors, had no doubt of it. Finally, for Ross, there was the added attraction of being the first man to reach both the south and north magnetic poles. According to the calculations of eminent German magnetic theorist Friedrich Gauss, the South Pole could be reached by sea, provided the huge crush of icebergs was avoided.

Britain had two formidable foreign rivals who shared these same aspirations and were already well ahead of Ross in their preparations. They were the aristocratic French botanist and explorer Captain Dumont d'Urville, capricious, haughty and brilliant, and the American naval lieutenant Charles Wilkes, a quarrelsome, arrogant but very capable astronomer-sailor.[7] Both had visited England in 1838 in preparation for mounting their major national expeditions south, not the least of their motives being to learn about the courses taken by two British whaling captains, James Weddell and John Briscoe, who had supposedly penetrated deep into the Antarctic icefields. Ross chafed at the political machinations and administrative vacillations that gave these two expeditions such a head start.

When Joseph Hooker hunted his captain down at the end of April 1839 to complain about his appointment, Ross was frantically trying to close the gap on his French and American rivals. Young Hooker might have been self-obsessed but he was no fool; he must have realised how close he was to angering his hard-pressed commander. Yet he could not stop himself. He was prepared to risk missing out on the voyage altogether rather than compromise on being the expedition's senior naturalist. What had led this lowly medical student to become so adamant and so unrealistic in his demands?

In defending the appointment of Robert McCormick to the expedition, Ross told Hooker that the veteran Scottish surgeon had been preparing for this southern voyage for three years. Hooker might have retorted that he'd been preparing for it for the whole of his life: he had wanted to be a naturalist-explorer as long as he could remember. Unlike the family of his hero and future friend Charles Darwin, the Hookers had actively encouraged naturalist exploration, not merely as a hobby, but as a stepping-stone to a possible career.

Naturalism could be said to have been absorbed with his mother's milk. Joseph's paternal and maternal grandfathers were both distinguished amateur botanists. Grandfather Joseph Hooker, after whom the boy was named, was a Norwich banker-merchant who in his spare time cultivated rare plants and also accrued a major entomological collection. On his mother's side, grandfather Dawson Turner was a Yarmouth banker with a passion for collecting non-flowering plants, on which he published several lavish illustrated volumes.

The painterly influence to support these botanical passions came from the maternal line on both sides: grandmother Hooker had been a Vincent, a family at the centre of the early-nineteenth-century Norwich school of landscape painters famously represented by John Crome and John Sell Cotman. Grandmother Turner, formerly Elizabeth Cotman, was a direct offspring of the same painting school. Both sides of the family collected paintings and encouraged sketching, painting and draughtsmanship in their children.[8]

Joseph's parents had amplified this heritage. His mother, Maria Turner, was a capable painter and draughtswoman taught by Cotman, who'd also been reared in a botanical household. Joseph's father, William Jackson Hooker, inherited a substantial endowment of land from a cousin, which subsidised his passionate hobby of botany. William had grown up roaming the woods and fields of Norfolk, collecting insects, lichens, ferns and, his particular favourite, mosses, of which he'd discovered a rare local species when he was twenty.

After befriending the haughty Yarmouth amateur botanist Dawson Turner and marrying his eldest daughter, William Hooker undertook a series of botanical collecting tours in 1807–8 among the rugged mountainous landscapes of Scotland and the Isles. His energy and ability caught the eye of Sir Joseph Banks, of Cook-voyage fame, who had since become the doyen of British botany and President of the Royal Society.

Sir Joseph funded William Hooker in 1809 to collect plants on his behalf in Iceland, a tour that turned into an adventure that featured storms, fires, shipwrecks and rebellions. On the voyage out, William befriended a talented adventurer named Jorgen Jorgensen, who saved him from a burning ship and then staged an audacious, one-man Icelandic revolution against the kingdom of Denmark. William's son Joseph would one day meet the same Jorgensen in Van Diemen's Land, where he'd been banished as a convict. Though William Hooker longed to pursue further botanical explorations – in Australia, Africa, Asia and South America – he was persuaded by his father-in-law to invest in and manage a brewery, a position for which he had neither inclination nor aptitude.

Confronted with a decade of business losses and an expanding family of two sons and three daughters, William was rescued from bankruptcy by the patronage of Banks, who in 1820 wangled him the newly created position of Professor of Botany at Glasgow University, even though William had never attended a university lecture in his life.

Becoming a professor saw William Jackson Hooker move down in the world. In early Victorian Britain, professors were generally poorly paid, with little cachet. They were expected to be independently wealthy and not to rely on paid teaching to make a living. Those who did so were thought rather vulgar. William Hooker had the additional disadvantage that his subject, botany, and his students, medical trainees, both ranked close to the bottom in university status. A few aristocratic amateur botanists, like Joseph Banks, might be revered within elite institutions like the Royal Society, but in general, botany was regarded as the preserve of hobbyists – women flower painters, self-taught artisans and minor clergy.

Botany's emphasis on the mechanical tasks of collecting, preserving, classifying and illustrating plants lacked the philosophical kudos of natural sciences such as astronomy, which prided itself on using inductive methods to investigate laws of causation within the cosmos. William's task at the university was humble: he was expected to provide a practical course for young doctors intending to work in the colonies, where local plants would often have to substitute for manufactured drugs. While William Hooker was not exactly poor, he and his large family always had to struggle to maintain their social standing, and William was reliant throughout his life on aristocratic patronage.

Hooker's social dependence scarred the thin-skinned Joseph deeply, creating in the boy a prickliness and a resentment that influenced the course of his life. Joseph worshipped his father and was acutely conscious of any slights directed towards his hero. The young Joseph even developed a strong dislike of his maternal grandfather, Dawson Turner, because the self-important banker was inclined to exploit the expertise of his son-in-law. It made Joseph burn with indignation.

As a father, William Hooker of Norwich was domineering but kindly, much in the mould of Dr Darwin of Shropshire. He resembled Dr Darwin, too, in being physically imposing, intelligent and fanatically hardworking. Yet Charles Darwin and his future best friend Joseph Hooker grew up in households whose ethos differed markedly from each other. It was not just that the Darwins were wealthier and socially superior, the family was also much more fun-loving. The Hookers, by contrast, drew on a flinty puritan heritage that had by the early nineteenth century mutated into devout Anglican evangelicalism. Joseph's parents were as austere, disciplined and moralistic as the Darwins were gossipy, convivial and sport-loving.

Perhaps this lack of sociability made the Hookers an exceptionally close-knit family. The five children had few friends and relied entirely on each other's company. They revered as much as adored their high-minded parents. Young Joseph was never allowed to sit down in his mother's presence unless given explicit permission. His father worked ceaselessly, churning out popular books and papers, as well as lectures, to subsidise the family income. As with the Darwins, the eldest Hooker son, William, was regarded as naturally brilliant but lazy; the younger, Joseph, as rather a 'plodder'. His mother described him as earnest and studious but not particularly bright. She admitted, however, that 'Joseph bends all his soul and spirit to the task', though even this underestimated the boy's ferocious stamina and powers of concentration.[9]

His parents failed to recognise his exceptional strength of character because the boy was physically slight and thought to be of delicate health. He was given the nickname 'croaky Joe', a relic of his childhood history of croup and scarlet fever. This contributed to the 'nervous irritability' that brought on heart palpitations whenever he had to speak in public.[10]

Notwithstanding any physical weakness, young Joe strained to imitate and impress his father as a field botanist, scorning toys, games and domesticity in favour of collecting insects and plants, especially mosses. Moss became a strange emblem for the boy. It had been finding a rare Norfolk moss that had launched his father's botanical passion, and William further stoked Joe's imagination by showing him a chunk of moss given him by Mungo Park. The great African explorer had supposedly gathered it as a rallying symbol in the Gobi Desert, when on the point of dying from thirst and exhaustion. Much later in life, Joseph Hooker was to describe himself as, by upbringing, 'a puppet of natural selection' and, by heredity, 'a born muscologist' – the technical term for someone who specialises in the study of mosses.[11]

While at Glasgow High School, where he studied mainly liberal arts and the classics, Joseph began, at the age of seven, to attend some of his father's university lectures. He was already well used to helping his father with his burgeoning private herbarium, an essential tool for any botanist with serious ambitions. Like William's other students, Joseph studied from a textbook written and illustrated by his father. Called *Flora Scotica*, this two-volume reference book with plates and descriptions of three hundred plants became a model of botanical scholarship that Joseph was to carry in his mind when journeying to the Southern Hemisphere.

Joseph also enjoyed listening to those of his father's students who carried a whiff of the exotic about them. Most of these were studying botany in preparation for working as doctors in the imperial lands of India, Africa and the Pacific. In such uncivilised places, botany mattered. A concrete knowledge of the pharmaceutical and economic value of plants could prove the difference between life and death. William had recognised the importance of using students as future providers and collectors of species and he consciously developed an expansive network of contacts around the empire – another inheritance that would prove invaluable to his son's future. Their pan-colonial samples of dried plants and flowers were added to William's growing herbarium, which functioned as a type of laboratory for students and colleagues to practise classifying rarer plant species.

In his lectures, William Hooker taught the two major prevalent systems of plant classification. The first was the longstanding and relatively easy Linnean or 'artificial' system, still widely followed in Britain in the 1830s. This used a simple method of counting a few of a plant's reproductive parts, such as stamens or pistils, then placing the plant in a hierarchy of groupings from the most general, called a kingdom, down through class, order and genus to the base category of species and its temporary varieties.

The second was the more complex and challenging system of 'natural' orders, which was gradually supplanting the Linnean system. Pioneered by European botanists Antoine-Laurent de Jussieu and Augustin-Pyramus de Candolle, it aspired to use all parts of the plant, including its structure, to determine classification. Under the influence of friends of William Hooker – Robert Brown, a former Australian voyager; and Dr Lindley, Professor of Botany at University College, London – the natural system had been extended to include minute dissections of seeds and other embryological characteristics. A knife and a microscope therefore became essential prerequisites for discovering natural affinities.[12]

Joseph Hooker's early letters suggest that he thought about little else but botany. His father described him as 'a zealous botanist', so committed that at the age of fifteen he preferred to stay home 'to study *orchidae* zealously' rather than take a holiday with his relatives. 'Earnest' and 'zealous' became Joseph's default labels, though they were matched by a less visible streak of stubbornness and bellicosity. On one occasion in May 1836, his mother reported that he'd got up early to walk the twenty-four miles from Helensburgh to Glasgow because the only ferry would have caused him to miss a lecture.[13] By this time, too, despite being by far the youngest student of his cohort, he was winning prizes for subjects like anatomy.

What leisure time he could snatch in between studying was spent building up a small herbarium of his own, or corresponding with his father's naturalist friends in other parts of Britain. On the surface, it would appear difficult to find two more dissimilar medical students than serious Joe Hooker and wild Charlie Darwin: they seemed unlikely future friends and collaborators.

Thankfully, Hooker did resemble Darwin in ways that softened some of the young botanist's otherwise daunting seriousness. Both shared a strong political liberalism, a whimsical sense of humour, a vivid pictorial sense, and a passion for the outdoors. Hooker, like Darwin, loved nothing more than tramping through rugged countryside, sleeping rough, hunting for rare specimens, and admiring the picturesque wonders of nature.

William Hooker organised regular field excursions for his Glasgow students. On Saturdays they stayed close to the town, usually travelling to Campsie Glen and Bowling Bay, but on more extended trips, taken annually in June in the western Highlands, around thirty students carried their own vasculums and plant-drying apparatus.

Avoiding tourist trails, William Hooker led the group across stony paths and overgrown tracks, using a highland pony cart to carry the marquee under which the students usually slept at night.

Young Joe never forgot the excitement of these excursions. They tramped around beautiful Loch Lomond, slept at the hamlet of Tarbet on heather spread on the floor of a crofter's cottage, and collected specimens as they went. Sometimes they tempered botanical work with other pleasures. Once, they continued on past Loch Lomond towards Killin, via the old anti-Jacobite road built by General Wade. On the way, they paused to fish in the fast-flowing trout streams and to scramble up the rocky slopes of the Breadalbane Mountains.

In 1836, Joseph, now aged eighteen, and a botanist friend, William Wilson, undertook a tour of their own. They climbed the Aberdeenshire mountains, and stayed in a rough cottage at the foot of Loch Lomond. The following year found Joseph geologising in Arran; the year after, he walked through Ireland accompanied by his father's botanical friend and mentor Dr Robert Graham.[14] Ireland was beautiful, but Scotland had captured his heart. The rugged, romantic western Highlands landscape became as much a part of Joseph Hooker's imagination as the mountains of North Wales were of Darwin's.

Joseph also shared with his future hero a boyish fantasy of voyaging to the far-flung ends of the globe. Like Darwin, he read *Boy's Own Paper* stories of explorers trying to penetrate the forbidden city of Timbuktu in Africa, or sailing among the coral islands of the South Seas. His father, who'd also once dreamed of voyaging to such places, earned some much-needed money and inflamed Joseph's fantasies by producing botanical drawings and plates for Captain Frederick Beechey's famous *Narrative of a Voyage to the Pacific and Beering's Strait* (1831), as well as *Niger Flora*, an account of west African plants.[15]

Grandfather Hooker had entranced Joseph from childhood by reading to him from his extensive library of travellers' tales. One image lingered always in the boy's mind: it was an illustration from James Cook's second voyage to the Pacific that showed sailors clubbing penguins for the pot. Behind them soared a rock arch overlooking Christmas Harbour, in Kerguelen's Land in the southern Indian Ocean.

Many years later, Hooker was to tell Darwin that 'from my earliest childhood I nourished and cherished the desire to make a creditable Journey in a new country, & with such a respectable account of its natural features, as should give me a niche among the scientific explorers of the globe'.[16] It was Darwin himself who crystallised this dream. In 1838, shortly after Joseph turned twenty, William Hooker lent his son the proof copy of a forthcoming book of naturalist exploration. It was Darwin's *Voyage of the Beagle*, the story of a naturalist's exploration of South America and the South Pacific, and one that embodied Joseph's keenest fantasies.

It all seemed so realistic, graphic and attainable. The voyage of the *Beagle* was only two years in the past, Darwin was only a dozen years older than Joseph, and he too had been a tyro naturalist when he embarked. This was not an account by some wild, eccentric explorer, it was the accessible journal of an ardent young British naturalist like himself. Joseph was so excited by the book that he slept with the proofs under his pillow, so he could rip through them in the early morning before lectures.

Though conscious of the range of Darwin's talents, Joseph announced to his recently knighted father, Sir William Hooker, that he intended 'to follow in Darwin's footsteps'. But lacking Darwin's private wealth, Joseph had only one practical option: he must join a naval expedition and voyage as a naturalist at the government's expense.

Sir William Hooker swung into action to support his son's plan, tapping the network of his senior scientific and naval friends to discover suitable expeditions projected for 1838–39. He had several times exercised his influence on behalf of students, so it was a well-oiled process. There turned out to be two possibilities: Captain Trotter, famed as a scourge of pirates, planned to go to the Niger to foster commercial treaties with native chiefs; Captain Ross, the famous polar explorer, was to lead a magnetic mission to the Antarctic. Fortunately for Joseph, Sir William was better placed, through his connections, to influence Ross, otherwise the boy might have died of malaria along with the rest of Trotter's men.

In the autumn of 1838, a family friend and neighbour, James Smith, arranged for Sir William and Joseph to meet with Captain Ross for a breakfast parley. Ross agreed to take Joseph along in some naturalist capacity, provided he completed the necessary qualifications to work as an assistant surgeon. Although Joseph had begun to prepare himself for such a possibility by taking courses at Glasgow University in anatomy, astronomy and navigation, he confessed in a letter to Grandfather Turner on 8 October that 'I can hardly conceive my being prepared both as a Medical Man and a Naturalist: to pass my necessary examinations will be a great push, while again if I do not devote a good part of this winter to Natural History, I had better not go at all.'[17]

For a time it looked as if Joseph's frenzied efforts to finish his degree would fail: the expedition was due to leave in May 1839, a few days before his qualifications were due to be completed. Again it was Sir William who came to the rescue: his pressure on Sir William Burnett, Physician General of the Navy, as well as on Sir John Barrow and Captain Ross himself, had 'the difficulty obviated' by obtaining permission for a late notification. In April 1839, another friend of Sir William, Dr John Richardson, who was in charge of prospective naval surgeons at the Royal Hospital in Haslar, offered Joseph a paid appointment until the Antarctic expedition was finalised.

Determined even now not to rest on his laurels, Joseph also began an intensive study of Van Diemen's Land plants. By early June, Sir William felt able to relax his lobbying efforts, confiding to Dawson Turner that Joseph's way was now clear to carve out 'an honourable scientific career'. Sir William hoped that, once his son returned from the expedition, 'if I have the same friends I have now, it may be in my power to keep this appointment [the Glasgow Professorship of Botany] in the family by applying to have it made over to Joseph'.[18]

No wonder, then, that Ross's announcement at the end of April that Joseph Hooker was to be not the expedition's naturalist but a flunkey to surgeon Robert McCormick proved so devastating. The intricate plans of father and son were shattered. It was a setback not unlike the one Darwin had faced when his father vetoed his going on the *Beagle* voyage.

But Joseph, again like Darwin, wasn't giving up without a fight. His spectacled earnestness belied a fearless spirit. He immediately began lobbying his father's influential naval and scientific friends, a formidable group that included Robert Brown at the British Museum; Sir John Barrow at the Admiralty; John Children, Secretary of the Royal Society; Edward Forster at the Linnean Society; Government Hydrographer, Francis Beaufort; and Vice President of the Royal Society, Sir John Lubbock. All of them, Joseph reported to his father, were outraged at his shabby treatment. Most 'disadvised me from going except as the only Naturalist on the ship'.[19]

When his first flush of anger subsided, however, Joseph realised he could not win. If he forced himself on a reluctant Captain Ross his position would obviously be untenable. Miserably he wrote to his father apologising for the waste of time, money and effort, adding that 'the most painful part of my duty remains to be done, viz, going to Captain Ross and respectfully declining the appointment'.[20]

To his relief, Joseph Hooker didn't have to resign after all. Captain Ross proved a model of diplomacy and went out of his way to reach a compromise with his aggrieved surgeon's mate, giving him the loose title of voyage 'botanist'. Joseph's official naval position did not change, but the captain gave him a private assurance that he'd be given every opportunity to go ashore, and eventually to publish his scientific findings. Ross's pliancy might have had something to do with Joseph's lobbying, but perhaps the captain was simply embarrassed because of his friendship with Sir William Hooker. Either way, he wrote urging his officers, especially surgeon McCormick, to make the young man welcome and to ensure that he had full access to botanical sites on the voyage.

McCormick, despite a reputation for pomposity and a bad temper, responded genially. As the senior officer, he could afford to be magnanimous, and having damaged his reputation during the *Beagle* voyage, he was no doubt anxious to please his captain. He proved kindness itself, even tending Hooker personally when the young man hurt himself boarding the unfamiliar ship. It was McCormick, too, who introduced Joseph to his newfound naturalist hero. The two doctors happened to bump into Charles Darwin when strolling together through Leicester Square. Joseph never forgot his first sight of Darwin's beetle-brow and quick, intelligent eyes.

McCormick also successfully petitioned the gunroom officers of the *Erebus* to allow young Hooker to share their mess. This unusual privilege brought him a superior level of comfort and space: a redwood- and maple-timbered room for recreation, and his own small cabin of painted satinwood with a bed, table, seat, drawers and a bookshelf.

During the five-month wait at Chatham docks while the *Erebus* and its companion ship the *Terror* were being refurbished and victualled, Hooker was inducted into the vagaries of the wooden world. The Admiralty had allocated the generous sum of one hundred thousand pounds for preparing the ships to cope with the Antarctic ice, and for the purchase of equipment to set up magnetic observatories. Hooker was impressed by the thoroughness with which the two former 'bomb ships' – already strongly built to withstand heavy mortar detonations – were transformed into the first dedicated British icebreakers.

> They were . . . trebled with linings outside of oak and bows filled with oak and African teak bound together with iron braces, till they . . . presented a mass of woodwork several foot through at the waterline. The copper sheathing used was of double thickness all round and quadrupled at the bows . . . great transverse beams crossed the vessel below the decks, the holds were divided into three watertight compartments by double walls of great thickness, and the space outside the ships where the channels run was filled up all along both for additional strength and to throw the ice off as the ship should surge into the water. The decks were further doubled, with the planks curiously dovetailed diagonally into one another.[21]

Besides this structural work, each ship was issued with nine support boats, numerous ice anchors, knives, chisels and ice saws, and enough stores and antiscorbutics to last the hundred and twenty men for at least five years. Most important of all, stoves were fitted in the bottom holds to funnel hot air along each side of the decks and through the officers' mess and cabins. The two icebreakers' only shortcoming, Hooker recalled, proved to be 'their poor sailing qualities, for manoeuvring amongst the ice'.[22]

The provisions made for zoology, geology, botany and maritime-biology work offered a sorry contrast. Admiral Parker of the Admiralty Board let it be known to Ross that natural history had nothing whatsoever to do with the expedition. As a result, McCormick told Joseph bitterly, the government was 'loathe to make grants for the natural history department'. The surgeon tried to rectify this as best he could from his own pocket.[23] The official issue for botanical work was laughable: it consisted of twenty-five reams of paper, two vascula, and two Ward's Cases for bringing back plants alive. Preserving jars and fluid were thought to be superfluous: empty pickle jars filled with naval rum had to serve as substitutes.

Fortunately the combined Hooker and Turner families provided some private help. William funded Joseph's uniform, personal stores and essential instruments, including a beautiful chronometer that survives today. He also kicked in fifty pounds worth of botanical and natural-history books. Dawson Turner gave his spiky grandson a fine travelling thermometer.

Naval officials were not the only scientific philistines. Joseph was disgusted by the ignorance and rudeness of the Commission of the Royal Society, appointed to convey broad scientific advice and policy to the expedition. He was summoned before their panel of scientists in mid-July to receive his botanical instructions. These, he told his father, consisted of 'much nonsense' about the different locations of botanical genera and a long list of platitudinous orders', after which he was dismissed without even being wished a successful voyage.[24] It was a discourtesy that the Royal Society would one day come to regret.

There is no evidence, however, that young Hooker experienced the same affliction of homesickness and doubt that depressed Darwin during his parallel experience of being stuck for months in limbo between shore and ship. Hooker filled his time by studying the plants of the Southern Hemisphere and learning the complicated techniques of magnetic observation.

One Sunday towards the end of August 1839, Sir William visited his son on the *Erebus* at Chatham dock to say farewell. Though warmly welcomed by Joseph's messmates, he was appalled at their salty language. Afterwards he wrote to his son expressing disappointment that the officers' conversation had not taken 'a more scientific and soberer turn'. He'd seen no signs among them of 'respect for the Sabbath'. It was important, he urged Joseph, 'to carry something of the Sabbath into the week. I am sure you will be a happier man for it.'[25]

Joseph made no response to this comment, a sign of what seems to have been his constitutional indifference to religion. He did, however, write to Sir William of his delight that all five mess companions shared his passion for reading voyages of exploration. The volumes of Cook's and Weddell's southern travels in the mess library were already well thumbed, and his new mates were queuing up to read Joseph's copy of Darwin's *Voyage of the Beagle*. Whether Sir William considered this acceptable fare for the Sabbath is not known.

Perhaps, along with the inevitable pangs of homesickness, Joseph felt a small flutter of relief at escaping his parents' evangelical scrutiny when, on 25 September 1839, the *Erebus* and *Terror* slipped their moorings at Gillingham's Reach and set sail down the Thames towards the open sea. A fortnight later, after encountering a frustrating succession of storms and contrary winds, they eventually cleared Lizard Point and headed for Madeira.

Ross summed up their feelings: 'It is not easy to describe the joy and lightheartedness we all felt as we passed the entrance of the Channel, bounding before a favourable breeze over the blue waves of the ocean, fairly embarked in the enterprise we had all so long desired to commence, and freed from the anxious and tedious operations of our protracted but requisite preparation.'[26]

# The Travails of a Young Botanist

On 14 November 1839, the *Erebus* and *Terror* anchored close to the harbour of Porto Praya on St Jago, where Darwin had had his revelatory moment staring at the cliff-face. Joseph Hooker thought Porto Praya a 'magnificent little town' and described the view with geological precision: 'On each side of the Bay, perpendicular, flat-topped cliffs . . . stretch along the coast, they are all of black trap rocks, sometimes columnar and about half way up, intersected by the broad white horizontal bands, which Mr Darwin described, and is the ancient sea beach of shell sand, resting upon one layer of lava and covered by another . . .'[1]

The ship had already called briefly at the islands of Madeira and Tenerife, where Hooker wrote lively accounts of jaunts among British émigrés which were reminiscent of Darwin's blend of poetry and science. Like Darwin, Hooker had begun his 'Antarctic Journal' as a daily conversation with his absent family. 'There are . . . peculiar emotions attending the seeing of new countries for the first time, which are quite indescribable,' he wrote on 17 March 1840 in an almost literal echo of Darwin. 'I never felt as I did on drawing near to Madeira and probably never shall again. Every knot that the ship approached called up new subjects of enquiry, and so it is with every new land and every barren rock.' He felt moved to file away these novel sensual and scientific impressions for later replay to those he loved: 'I never view a new scene but I think what pleasure it will give me to view with you all, to map the places where my specimens were gathered, to paint the views to my mother and to spin to William the yarns and incidents that befell my excursions, while my grandpa and my sisters will look upon me as "the Monkey that has seen the world".'[2]

For the next seven days, while Ross took on fresh supplies and made magnetic observations, Joseph Hooker set out to walk Darwin's trails. His immediate thought was to take the same tracks to the west of the town that had led Darwin to a giant baobab tree, introduced from Africa by the Portuguese many years before. As he and the ship's purser set off along the beach, with the temperature at 84° F in the shade, Hooker echoed Darwin's thrill at first encountering tropical flora: 'we descended at the other end of the Town . . . and came to a beautiful grove of Cocoanuts, bananas, Palmettos etc. . . . being the first I had seen, they delighted me extremely. Those who have not been in a tropical climate can have no idea of the graceful beauty of this area . . .' On the way they passed by the exact stand of stunted acacias twisted into strange shapes by the wind that Darwin had recorded. A poor black family, seeing the two Englishmen struggling like mad dogs in the noonday sun, generously gave them 'some of the most magnificent oranges I ever ate, of an immense size, and very sweet'.[3]

Somehow, though, they missed the path to the baobab tree, forcing Hooker to reattempt the excursion next day with surgeon McCormick, who'd walked with Darwin to the site seven years previously. At that time, McCormick had scrambled up the bulbous grey trunk to inscribe his initials and the date, 1832. In Hooker's company, he now added 1839. Both men then sketched the strange pachydermous tree, which Hooker assessed at around fifty foot in height and thirty-eight in girth. The young botanist was able to pull down one of its pendulous six-inch seedpods, which later ripened on the ship and carried the sickly-sweet scent of tropical Africa through his cabin.[4]

With Darwin's journal as a guide, Hooker made several excursions inland to St Domingo, his eyes attentive as much to the geology as the botany. He noticed how rocks were coated with crystallised carbonate of lime, a metamorphosis that Darwin had described 'beautifully'. He walked twelve miles through a steep valley, over boulders and water-worn pebbles that scorched his feet, until he reached the decrepit town of St Domingo. Darwin had wryly dubbed its sorry-looking houses and decaying church the '*Ribiera grande*'. On a second excursion, Hooker found time to stuff his pockets with plants, including a bright orange-flowered Compositae and a beautiful blue campanula.

As he clambered up the steep cliffs to the top of the mountain behind the town, he found himself having to ransack his memory for literary analogies to describe the grandeur of the scene below. It reminded him of *The Arabian Nights* and of Dr Johnson's setting for the romantic story of Rasselas, the melancholy prince of Abyssinia who journeyed fruitlessly to Egypt in search of enlightenment.[5]

On 18 November, the afternoon before they were due to sail, Hooker found time to make an excursion to nearby Quail Island, where Darwin experienced the epiphany that decided him to write on the geology of his travels. Following the same paths, Hooker gathered around twenty species of plants. He guessed that they'd migrated, like the rest of the Cape Verde flora, from the south of Europe, the African desert and tropical Africa, but he had no idea how.

Surprisingly, though, he spent relatively little time on botanical collecting. It was clear that at this early point in the voyage, he still regarded himself as a broad-ranging naturalist in Darwin's mould. On the water's edge he hunted for, but failed to find, signs of the octopus that Darwin had described. His keenest concern was to test his own powers of geological observation against Darwin's. To his delight, he came up with a 'likely theory' that his hero had missed.

In between the hardened ash of the beach and a recent lava flow, Hooker noticed a bed of disintegrating volcanic material full of pebbles but devoid of shells: he decided it must be 'a mass of matter carried forth by the under lava stream, and which from being above the surface of the sea, or tide, during the formation of the neighbouring sea beach, was deprived of sea shells and other fossil remains'.[6] The observation was insignificant, but it made Hooker feel like an established Lyellian naturalist.

Hooker had a dilemma when it came to using Darwin's examples in theorising. He couldn't help treating *Voyage of the Beagle* as an intimate 'companion'. Where else would he find such a sympathetic mind, such an evocative writer, such a luminous observer and theorist? The book had been a key reason for his decision to embark on a naturalist voyage, as well as the inspiration to write his own journal. Darwin, too, was the model that Captain Ross had held out to him as a shipboard naturalist. At one level, then, it seemed obvious for Hooker to cruise in Darwin's wake – psychologically, if not always literally – to indulge in a feeling of kinship with him, as Darwin had done with Humboldt.

But Hooker was also a proud, ambitious young man and he wanted to make his own ripples in that world – no easy matter when Darwin had possessed such advantages of opportunity and ability. At Saint Paul's Rocks, their next destination after Cape Verde, Hooker, as a lowly surgeon's mate, didn't even get the chance to make an independent assessment of the biota on this scattering of rocks five hundred miles east of America. While Ross was attempting to land the first shore party in the pinnace, one of the sailors was tossed into the sea by the wild surf, to be menaced by sharks, after which the captain banned any further landings. Aside from seeing a seaweed that McCormick brought back, Hooker had to rely entirely on Darwin's 'admirable description' of the odd bird, crab and insect that he'd collected there with FitzRoy.[7]

On 31 January 1840, after seventy-four days of further sailing, they reached the island fortress of St Helena and anchored in Jamestown harbour, where, Hooker noted, 'guns are pointed from every nook and corner' and 'batteries bristle everywhere'.[8] Famed as the site of the exile, death and entombment of Napoleon Bonaparte, the fifty-square-mile island presented Hooker with his first real opportunity to struggle out from under Darwin's shadow. The *Beagle* had called there briefly at the end of its long voyage, when Darwin had only England on his mind, so his journal had been brief. Captain Ross, however, had been charged to erect a permanent observatory there, which would take at least a week or more.

Hooker knew that the history and environment of this remote southern landmass, nearly two thousand miles from America and twelve hundred miles from Africa, had intrigued botanists for several centuries. A succession of European settlements, beginning with the Portuguese in the sixteenth century and culminating in annexation by the British in 1834, meant that the island's flora had been subjected to waves of human depredation and imported plantings. Since much of this imposed flora was British – reminding Hooker intensely of his beloved Scottish Highlands – he could be more than usually confident about his identifications and taxonomies.

Better still, the island presented a perfect case study for exploring a variety of botanical issues that had always intrigued him. Hooker nursed an ambition to raise the discipline of botany to greater respectability by working out some of its little-known causal puzzles.

Islands produced 'insular floras', and insular floras made fascinating study. How had plants been dispersed from their original habitats? How was it that introduced species could compete with, and often eclipse, indigenous ones, despite the fact that the latter had supposedly been created expressly to suit local conditions? And how had the extinction of local flora affected the larger natural economy of insects, birds and animals?

Many years later, Hooker was to declare that 'Oceanic islands are in fact, to the naturalist, what comets and meteorites are to the astronomer.'[9] They were natural laboratories. Joseph Banks and a string of naturalists after him had commented on the ravages of introduced species at St Helena. Darwin himself had been able to locate only fifty-two indigenous plant species, compared with 424 imported ones. Hooker intended to produce a more systematic botanical analysis of this 'very interesting island', which would mean a careful mapping of all species in relation to environmental factors, such as elevation, moisture and soil type.[10]

During the next six days, he made excursions all over St Helena in company with helpful locals and shipmates. They first forced him to visit the island's most famous tourist site, Napoleon's tomb. Revolted by the hard sell of 'the mighty dead', Hooker took to his heels, 'heartily wishing for my own sake as well as for the good cause of humanity, the Emperor had his wish of living and dying in some remote corner of Britain'.[11]

For the remainder of the time ashore, he concentrated on collecting plant specimens. This meant climbing seven thousand feet to the top of the highest mountain, Diana's Peak, then traversing 'precipitous valleys' in all directions. Whatever specimens he couldn't dig up, he sketched and described in detail, including his first sight of a gigantic tree fern. As well, he tested the moisture levels of the air and soil with his hygrometer, and examined how British, African and European plants had adapted to the distinctive climatic conditions of the two sides of the island, one as barren and dry as a desert, the other as wet and misty as Scotland. Humboldt-style, he relied 'chiefly on my own resources' to map the growth of native and indigenous vegetation within a series of altitude zones, ranging from 'maritime' seaweeds to mosses growing in alpine heights.[12]

By the time they sailed for Cape Town on 10 February, Joseph Hooker was pleased with himself. He'd proved in detail that introduced species from Europe were killing off existing populations of forest trees, most of which had originally migrated from Africa. Furthermore he'd observed that these foreign imports had altered the local soil composition and microclimates to such an extent that the indigenous species could never return. Neither could the many other interdependent plants, insects and animals that had also disappeared.[13]

It was clear to Hooker that island species were especially vulnerable to degradation and extinction. All this cast doubt on orthodox creationist ideas that species had been exclusively designed for particular climates. At the very least the island species must have migrated from some other centre of creation, and adapted very effectively to new conditions. Moreover this was a perpetual process of invasion and struggle.

His finest trophy was the discovery of a rare Brazilian *Araucaria* tree. He described it in minute detail and also disproved previous 'travellers' testimonies' by demonstrating that it was climbable. Nimble and skinny, Hooker swarmed up the scaly bark to knock down several spiky six-inch cones. He was so sure the *Araucaria* would 'prove a nearly unknown tree' that he displayed the cones in his cabin, tied up with 'lover's knots'.[14]

Though his other botanical samples were less exotic, he was satisfied that within the limitations of the late summer season he'd made a sound collection to send to his father. He actually dispatched four sets to England, along with a backup set to Cape

Town, urging the need for discretion because he was technically flouting McCormick's (and hence the Admiralty's) monopoly on voyage collections. Having to procure such a wide representation of flora in such a short time had shown him the importance of relying on the expertise and goodwill of local enthusiasts. He thanked a string of St Helena merchants, administrators, soldiers, and a young amateur woman collector, all of whom had generously supplied him with information and samples.[15]

His collections led to a surge of self-confidence that tempted Hooker to measure his achievements against Darwin's. In letters to his father from St Helena and Cape Town, he ventured a dig at his hero by scorning naturalists who complained of the tedium of voyaging and the miseries of being seasick, 'which I have never been even for an hour'.[16] He began to feel, too, that he was attaining a scientific status on the *Erebus* comparable to Darwin's on the *Beagle*. He boasted in letters home that Ross had put him in sole charge of the ship's tow nets, and had asked him to monitor their rich daily catches of open-water sea creatures for inclusion in the captain's private collection. Ross so admired Hooker's sketching abilities that he made him spend the bulk of his time dissecting and drawing this daily zoological haul.

Having already completed some hundred drawings, Hooker felt proud to have been given responsibility for 'a new field which none but an artist could prosecute at sea'. As he told his father excitedly, 'the number of species of little winged and footed shells provided with wings, sails, bladders and swimmers appears quite marvellous'. He'd also made some marine discoveries of his own: the strange luminous patches in the ocean usually attributed to electricity or phosphorous were actually produced by tiny animals, *Entomostraca crustacea*, which Hooker detected under Ross's microscope. Still more important, he'd found the presence of minute and unexpectedly intricate animal life called protozoa at the extreme depth of four hundred fathoms. These protozoa, he surmised, were staple food for fish and other marine creatures.[17]

As a result of all this work, Ross treated Hooker way above his formal rank. The captain was 'polite and attentive' and went to exceptional lengths to encourage their joint zoological work. He'd made space in his private cabin for Hooker to dissect, draw and store the daily catch. He was also willing to share his considerable knowledge of the lower orders of marine species and to allow generous use of his prized compound microscope. Sitting at the table each evening in Ross's company made the hours pass 'quickly and pleasantly', Hooker declared. 'It would have amused you,' he wrote home, 'to have come into the cabin and seen the Captain and myself with our sleeves tucked up picking seaweed roots and depositing the treasures . . . in salt water.'[18]

Naturally Hooker felt bound to reciprocate, by volunteering to make hygrometer readings of the atmospheric moisture at least four times a day, and twice a week at three am, as well as assisting Ross with experiments to take deep-sea soundings and sea temperatures. He felt he was helping the distinguished polar explorer to build up a new science of the ocean.

Hooker's fears of alienating McCormick, as Darwin had done on the *Beagle* voyage, were also proving groundless. The surgeon was interested only in collecting specimens of rocks and birds, and was happy to leave all the botany and other zoology entirely in his young assistant's hands. 'McCormick and I are exceedingly good friends and no jealousy exists between us regarding my taking most of his dept,' Hooker reported smugly. '. . . I am, *nolens volens*, the Naturalist'.[19]

On the strength of his early achievements, Hooker felt qualified to call himself a naturalist. Naïvely he wrote to tell his father that he'd decided to leave the navy at the end of the voyage to work as a botanist. Given his dedication, Hooker felt certain he would succeed: 'Gaiety has still less charms for me. If ever, on my return, I am enabled to follow up botany ashore, I shall live the life of a hermit, as far as Society is concerned.'[20]

So great was the time gap before he received letters from home – sometimes more than a year – that Joseph Hooker had no idea his cockiness was jarringly out of step with the mood of his family.

For a start, his mother and father were fretting about his health. With his tendency to bronchial problems and heart palpitations, it was not surprising that the Hookers regarded Joseph as fragile. One of his first letters home had increased their anxiety by revealing that he'd contracted a dose of rheumatic fever from lying on wet grass at Madeira. Though Hooker himself made light of it – he'd only stayed in bed a week – one of his well-meaning messmates subsequently sent an alarmist letter to William Hooker from Tenerife. It described their youngest son as 'in a very delicate state of health, looking wretchedly ill and thin', and concluded that he was 'quite unfit to undergo the fatigues of such a voyage'.

William dashed off worried letters to McCormick and Ross. Without consulting Joseph, he asked the captain to leave the boy behind in Van Diemen's Land or New Zealand, instead of exposing him 'to the difficulties and privations for which his constitution is unsuited'. Fortunately the letter didn't reach the ship for a further six months, by which time the family had received independent assurances that their son was in the best health of his life, a point that the captain was able to reaffirm in a letter written from Van Diemen's Land.[21]

The Hookers' parental anxiety was understandable: three of Joseph's siblings had fallen ill since he'd left for the southern oceans. William Hooker, Joseph's older brother and best friend, had developed ominous signs of consumption, which soon also appeared in two of his sisters, Elizabeth and Mary Harriette. William, recently married and with a young child, decided to improve his health by taking up a medical position in Jamaica, while the two girls were sent to a clinic in Leamington for treatment. The erratic mail meant that for the eleven-month voyage to their southern base of Hobart, Joseph had no knowledge of the plight of 'the seven people that I love best in the world'.

Hearing at Cape Town of William's recent move to the West Indies, and unaware of the seriousness of his illness, he wrote back to his parents nostalgically: 'if you but knew how often I think and dream of him, you would not be surprised at the sorrow I felt that he should be parted from you, though doubtless it is for the best'.[22] By then William was already dead of yellow fever.

Joseph Hooker's first intimation of this tragedy came when he grabbed a bundle of fifteen letters waiting for him in Hobart in August 1840. A black-edged envelope, addressed in his father's hand, caught his eye. He ripped it open to read the sickening words, 'My dear and only son'. While his shipmates gossiped about news from home, Hooker sat in silent misery. He remembered his brother as 'so warm-hearted a fellow that he would cut his right hand off to help a stranger'. All the anticipated pleasure of returning with dazzling tales of adventure turned sour.

'Now he is gone,' Hooker wrote bleakly to his mother, '. . . there will be none of my childhood's playmates when I return to talk over bygone times with, for he was at school my only companion'. His parents had 'lost a son, but I a brother and companion of 20 years' standing'. Another letter from his father in the same bundle turned the knife in the wound by revealing that his two sisters were also gravely ill. Though Elizabeth eventually recovered, his baby sister Mary wasted away to die around a year later.[23]

Further letters in the bundle contained a shock of a different kind. Sir William and his botanical friends, especially the famous Australian floral expert Robert Brown, expressed 'considerable disappointment' with Joseph's samples from Madeira, Cape Verde Island and St Helena. Part of the problem was his inexperience at working in tropical conditions, but he was also, his father implied, devoting too much time ashore on Darwin-style social excursions, and, at sea, on Captain Ross's maritime zoology.

The latter was a particular worry because it might woo Joseph away from the less-prestigious vocation of botany. William Hooker's dearest wish was for his son to inherit his books, herbaria and job in Glasgow. A young man without an independent fortune could not afford to behave like a dilettante. It was all very well for Darwin – he was a monied gentleman who could indulge his wide scientific curiosity and spend time gathering colourful experiences for a travel book. Joseph could afford no such luxury: his only chance of obtaining one of the rare paid university positions in Britain, such as the Glasgow lectureship, was to devote all his energies to botany. Judging by the samples William had been sent so far, Joseph had a long way to go. The collections were skimpy, poorly preserved and unsystematic.

Another of Sir William's botanist friends in Glasgow – no doubt nudged by Sir William – had also sent a letter to Hobart. It rubbed salt in the wound by telling Joseph that the naturalist Hugh Cuming had visited St Helena soon after him and managed in one day to gather a better collection of flora than he'd accrued in a week.[24]

William Hooker made it clear that he wanted his son to produce botanical work that would establish a reputation. This meant being more systematic and focused. He should aim at collecting comprehensive samples from some of the less-explored southern countries, and then publish detailed descriptions and sketches along the lines of William's own *Flora Scotica* – a work valued both for scientific reasons and for the economic potential of the plants it included. This was the advice given Sir William by the great Joseph Banks. Joseph Hooker's expressed plan to produce a flora of Antarctic plants would be interesting, certainly, but necessarily limited because of the scarcity of plant life there. He should widen his horizons to encompass the larger floras of Van Diemen's Land, New Zealand, and the like, which were relatively unexplored.

It was true, Sir William conceded, that it was difficult to devote time to collecting as well as to the intellectual work of identifying, dissecting, describing and sketching specimens, but there were ways to do both. Joseph should prevail on his shipboard friends and local experts to help gather specimens, or perhaps even hire an assistant – his father was willing to meet the costs.

As well, Sir William advised his son to develop specialised expertise in some southern genera, as he himself had done with Northern Hemisphere ferns. He suggested non-flowering plants, cryptogams, given that Joseph was eventually bound for Antarctica, where these were likely to predominate. In Van Diemen's Land, too, he should visit some of the high mountains, 'which everywhere afford what I consider by far the most interesting plants'.[25]

Above all, he needed to collect both more widely and more deeply, gathering specimens 'in *quantities* not in driblets', neglecting nothing, because 'such scraps as are useless for other purposes may yet, so long as they exhibit the Natural Order to which they belong, prove of service in illustrating the geography of plants'. Did Joseph not realise the precious opportunity that had been delivered to him? No foreign botanist had visited southern New Zealand since the time of Captain Vancouver in the 1790s. Was he going to squander his time on trivial shore entertainments, or use it to sketch and dissect molluscs? William Hooker admitted sorrowfully that he had not dared tell British Museum botanist Robert Brown that Joseph had been wasting his time on maritime zoology: 'Brown's idea is that . . . your time even at sea ought to be *mainly* devoted to studying the plants you have collected'.[26]

Poor Joseph Hooker. Devastated by the death of his brother, he now also had to swallow the bitter pill of having failed the man whose good opinion he valued most in the world. Emotionally tough though he was, Hooker's confidence in his botanical capacities was shaken.

Still, he tried to defend himself against the less reasonable criticisms. Did his father realise that he'd visited St Helena in the height of summer during a severe drought, when most of the plants were withered even before he picked them? Working at sea was also a deal more difficult than Sir William might appreciate. He had to spread out his collections in the middle of the sick bay, or on the gunroom floor where all his colleagues tramped and where cockroaches swarmed. Such was the humidity on board that it was almost impossible to dry specimens without damaging them.[27]

Adding to Hooker's travails, the pitching motion of the ship and his weak eyes made microscopy and accurate sketching difficult. Essential reference books were almost nonexistent. And his time was not his own: assistant surgeon Joseph Hooker was an officer in a naval command, answerable to the demands of his captain. And there were demands aplenty.

Perhaps most galling of all, the excruciating slowness of the mail meant that Joseph's achievements were being judged on out-of-date samples. William Hooker and his botanist friends had not yet, for example, received any specimens or reports from the site that Joseph was later to call 'one of the most remarkable islands on the face of the globe, remarkable alike for its isolation and the peculiarity of the few productions it boasts of'.[28] He was referring to Kerguelen's Land, a lonely archipelago deep in the southern Indian

Ocean, halfway between Africa and Australia. The *Erebus* and *Terror* had reached there from Cape Town in early May 1840, with a mission to stay for two months in midwinter while Ross set up magnetic and astronomical observatories.

The archipelago had originally been named the Isles of Fortune by its first European discoverer, Frenchman Yves-Joseph de Kerguelen-Trémarec, but on calling there in 1776, the blunt Yorkshireman James Cook had changed the name to the Islands of Desolation. The crew of the *Erebus* experienced an immediate taste of what Cook meant when it took four days of beating through storms to get close enough to the island to warp the ships into the only safe anchorage, at Christmas Harbour. Experienced seaman Sergeant Cunningham of the *Terror* doubted whether even this spot was reliable: it seemed to him 'the most dangerous place I ever was in, there is breakers all around and one reef runs out nearly a mile with a tremendous surf breaking over it'.[29]

If anything, Cunningham was understating. North-westerly gales blew for forty-five out of the sixty-eight days of their stay, and only three of those days were free from snow showers. And what gales! More often than not the sailors had to take their watches lying flat on the deck to avoid being blown overboard. On one occasion, Hooker recorded, the ship drove towards the rocks when all three of their anchors were set on a hundred and fifty fathoms of heavy chain: 'such a thing was never heard of before'.[30]

Only to Joseph Hooker was this archipelago a paradise. The harbour, with its distinctive arched rock, had been a keystone of his imagination since he'd sat on Grandfather Hooker's knee to stare at an engraving of Cook's crew bashing penguins for the pot. Fittingly, one of Hooker's first sights when walking along 'the beach of black sand described by Cook . . . were a few Penguins, droll looking creatures with black backs and white breasts like pinafores, their flippers hanging down like swords from their sides'. Soon he was helping his shipmates dispatch them to make a soup, though the dark flesh proved too rich for his taste.

The scenery around him was also as starkly romantic as he'd imagined: 'wild and desolate in the extreme; the mountains, which shut out my view on every side, were half covered with snow; the cataracts were frozen and not a ripple agitated the waters of the bay, which reflected the black cliffs like a mirror. Nothing but the scream of a gull broke the silence.' Where Cook's naturalist, Mr Anderson, had complained of being able to find only a handful of plants, Hooker saw nothing but botanical promise. The broad bands of vivid green and brown that streaked the rocks seemed 'sure indications of abundance of mosses'.[31]

So it proved. There were abundant mosses, and the snow-strewn ground was rich in every type of non-flowering plant, including lichens and seaweeds. Hooker set about investigating these little-known genera in detail, anticipating by several months his father's urging that he develop some botanical speciality. On the day of 16 May alone he gathered ten new species of lichen, including a magnificent specimen of the genus *Coloniza* with branching sulphur-yellow stems that were accentuated by a jet-black *Apothecia* nearby.

Not that these specimens were easy to come by. Sometimes he had to clamber down steep cliffs covered in ice and snow, and 'many of my best little lichens were gathered by hammering out the tufts or sitting on them until they thawed'. Most had to be chipped off with chunks of stone still attached.

Seaweeds were almost as prolific: one type formed a thirty-yard belt around the margins of the bay, a few yards from the shore. Its long leathery fronds created a 'submarine forest', perfect for sheltering crustaceans but difficult to row a boat through. Another type, the *Laminaria*, drooped down from jagged rocks on the edge of the deep water. Though lashed 'backwards and forwards by the swell and reflux with a deafening roar', the seaweed stoutly defied the heaviest surf.[32] Often the driving sleet and snow was so thick that Hooker could barely see a foot in front of him. On one occasion he took the *Erebus*'s sailing pinnace to gather seaweed floating on the water, but the boat was dismasted and nearly sunk in a gale. Ross immediately banned all further sea excursions.

Joseph Hooker wasn't exaggerating, though he was trying to prove a point, when he later wrote to his father saying that while in Kerguelen's Land he'd attended 'to nothing but botany' and had 'strained every nerve' to collect cryptogams.[33] Despite conditions that kept most of the crewmen cowering below decks, he managed to collect a hundred and fifty botanical species, more than doubling the known flora of the archipelago.

And these were by no means his only achievements: it's difficult to imagine any modern ecological survey, with all of today's comforts and technological advantages, producing a more searching report than his. As in St Helena, he mapped the incidence of all the island's botanical species along a grid of elevation, and he described and sketched a large proportion of the individual plants. He discovered a bizarre aquatic plant in a small lake, the fruit of which matured inches beneath the surface of the icy water. He found fossilised trees encased in rocks, which the geologist McCormick had overlooked. These, and large seams of coal that Hooker also uncovered, suggested that the archipelago had once been densely forested.

Hooker also attended to one of his father's key interests: species that might have an economic benefit back home in Britain. With this in mind, he produced the first thorough botanical description of the Kerguelen's Land cabbage plant. Mentioned by Cook, this indigenous plant had long intrigued botanists like Robert Brown. As well as speculating on its taxonomy and testing its famed antiscorbutic properties, Hooker reported that the roots tasted like horseradish, mustard and cress, and the boiled leaves like bitter cabbage. Such a useful plant, he thought, 'seems expressly planted by nature's hand for the poor mariner, when suffering under his own peculiar malady'.[34]

He gathered numerous seed samples, which he germinated easily under the stern window of Ross's cabin. Later he was able to give his father a small dose of his own medicine, when Sir William and his staff at Kew Gardens (where he became director in 1841) consistently failed to germinate the cabbage plant seeds Joseph had sent them. He implied mischievously that they simply weren't trying hard enough.

He also began to make tentative forays into 'philosophical botany', by which he meant moving beyond the usual taxonomic concerns of classifying plants in search of larger, law-like patterns, such as those that governed the origins, geographical dispersals and adaptations of plant types. He theorised that the climates of oceanic islands, even those as extreme as Kerguelen's Land, had led to a fascinating series of adaptations. The lack of seasonal variation on that archipelago had induced the plants to become perennials: large numbers were in flower even in midwinter, and few seemed troubled by ice and snow.

Hooker further speculated that the fossil evidence of previous forestation probably derived from a period 'when the Island may have enjoyed a very different geographical position with regard to neighbouring lands than it now does'. Although neither he nor anyone else understood the science of tectonic plates, he did think it likely that changes in ocean and seabed levels were likely to have brought new portions of land into being and sunk others from sight, causing land bridges along which plants had travelled to disappear.[35]

While island plants usually shared the character of those on the nearest continent, Hooker believed 'the vegetation here is peculiar to itself'. Kerguelen's Land was too far off the beaten track for birds or ocean currents from Australia or Africa to deposit foreign seeds. At the same time it struck him as remarkable that there should be such a close analogy between the Kerguelen's Land vegetation and that of the Arctic region, despite their being 'on two opposite parts of the world [and] when it is considered the vegetation of one is borrowed and on the other peculiar and limited to two degrees of Latitude and one of Longitude'.

Almost as surprising was the resemblance between Kerguelen's Land and American plants, while little or nothing had come from Africa or Australia, which were much closer geographically. Since Hooker believed that all species of plants and animals had come into existence only once and probably at a single centre, it was clear that by some means or other Kerguelen's Land plants had travelled and 'assimilated themselves to' the distinctive variations of temperature and soil in this isolated southern place. But how had this happened?

Joseph Hooker simply did not know.[36] Neither did he know that on the other side of the world, Charles Darwin was puzzling over similar problems, which he would before long be asking Hooker to help him solve.

Needless to say, when the expedition finally left the archipelago on 20 July 1840, on the last leg of the eleven-month journey to their intended southern base of Hobart, the crew could not contain their relief and joy – all except Hooker, who was 'really sorry to leave . . . it was not without a sigh that I saw the terraced shores of the Island of Desolation . . . sink step by step below the horizon till the last band of snow and the last black peak had disappeared in the waters of the Polar Ocean'.

Kerguelen's Land had been for him what Quail Island was for Darwin: 'To the last day I went botanising,' he later reported to his father. 'You cannot conceive the delight which the new discoveries afforded as they slowly revealed themselves.' Hooker had also come to appreciate what Darwin had found when visiting other stark and loveless places like Patagonia and the Galapagos Islands: 'by finding food for the mind one may grow attached to the most wretched spots on the globe'.[37]

Though they quickly put Kerguelen's Land behind them, its 'wretched' weather stayed with the ships. The day after their departure, the barometer plunged to its lowest, denoting that the gale had 'increased to nearly a hurricane'.[38] No sooner had the *Erebus* battled through that, losing contact with the *Terror* in the process, than Hooker heard an anguished cry of 'Man overboard.' It was the genial boatswain, Mr Roberts, who'd been kneeling in the bulwarks when the main staysail sheet suddenly whipped up and pitched him into the sea.

Through the gloom they could see a tiny figure struggling in the violent waves. Two boats were dispatched but one capsized in the process, washing two more sailors into the water. Both managed to grab onto a painter that the *Erebus* happened to be towing at the time, but Roberts was not so lucky. Hooker never forgot 'the fearful . . . sight of a fellow creature struggling for his life in so mighty an element as the open ocean', while overhead an albatross, 'already marking him for her prey, sweeps round and round the victim . . . [and] attacks the drowning man with both bill and claws'.[39] Only his hat was recovered.

Gales, rain, sleet and hailstones – the largest Hooker had ever seen – followed them for the remainder of the voyage. On 13 August, with the South West Cape of Van Diemen's Land already in sight, the barometer plunged once again. This time the gale was so severe it carried away the triple-reefed fore topsail, then shredded three consecutive main staysails and the fore staysail. As Captain Ross recorded in his journal, 'no canvass could stand up against such a storm'.[40] Several boats on the foredeck were flooded by the seas, and to Hooker's chagrin, 'many of my plants and Ward's cases were severely damaged'. Here was another setback that Sir William would never appreciate.

Driven miles off course by the storm, it took a further three days for the *Erebus* to beat back towards their original position. Then at last, on 15 August 1840, by the light of a bright full moon, they saw the coast of Van Diemen's Land directly ahead. Early the next morning, they entered Storm Bay and anchored off the Iron Pot lighthouse, at the entrance of the Derwent River. 'Nothing could exceed the pleasure the land covered with wood gave us,' wrote Joseph Hooker.[41] To an aspiring botanist, there could be no more beautiful sight.

R. McCormick, R.N., del.
Vincent Brooks, Day & Son, Lith.
The Baobab Tree of Porta Praya, St. Jago, Cape de Verde Islands. The trunk 36½ feet in circumference.
PAGE 16—VOL. I.


*The baobab tree, St Jago, Cape Verde Islands*

# Pilgrims and Pioneers

When the *Erebus* and *Terror* docked in Hobart on 16 August 1840, Captain James Clark Ross and his expedition were greeted by dispiriting news. The press in Van Diemen's Land revealed that both the French and American expeditions had 'forestalled' them twelve months earlier by adopting Ross's pre-announced route into Antarctica. Captain Dumont D'Urville had already discovered a new stretch of territory, which he named Terre Adelie, after his wife, and the American explorer Charles Wilkes was also claiming to have sighted chunks of a southern continent. Neither expedition, however, had actually made landfall, or reached the south magnetic pole.

Captain Ross decided to flaunt his displeasure by planning a more easterly route into Antarctica along the 170th meridian, close to latitude 69°S, where a British sealer had once reported seeing open water beyond the pack ice.

It was some consolation that Ross's expedition received a heroes' welcome from the people of Hobart, even before setting out on their first attempt on the pole. Sir John Franklin, the tubby, amiable Governor of Van Diemen's Land, who was Ross's old friend and a fellow polar explorer, welcomed them exuberantly. Famed for having eaten his leather boots while seeking the North-west Passage, Franklin was finding governing the convict colony less digestible. The snobbery, petty intrigues and malice of the dominant settler faction bewildered and distressed him. Among other things, locals called him an 'imbecile' and his vigorous younger wife, Lady Jane Franklin, an interfering bluestocking who wore the trousers. No wonder the Franklins were delighted to see British officers like Ross and Crozier, 'the excellent of the earth and the friends of our hearts'.[1]

The expedition's men blazed like comets in the isolated little settlement of 42 000 souls, nearly half of whom were convicts. There were also a few hundred beleaguered Aborigines. Sir John, 'bustling and frisky and merry with his new companions', threw himself into Ross's mission to establish a magnetic observatory.[2] Within a whirlwind nine days, he had organised two hundred convicts to erect a stone observatory. This he called Rossbank, to match the expedition's newly named anchorage of Ross Cove.

Not to be outdone, Lady Franklin commissioned a portrait of Franklin, Crozier and Ross standing imposingly outside the observatory, the last wearing a 'dear bunch of wattle in your buttonhole' that she'd given him as a memento of Van Diemen's Land.[3] She also plied Ross with pots of homemade marmalade for the forthcoming voyage. The crew found different sorts of pleasure drinking at a pub called the Sign of the Gordon, and eating fresh vegetables supplied from the governor's garden.[4]

Joseph Hooker's excitement derived less from the beauty of the landscape or the hospitality of the inhabitants – much as he appreciated both – than from the unexpected opportunity that Van Diemen's Land provided for researching the flora of 'the most interesting islands in the Southern Hemisphere'.[5] Though he still doubted his capacity to undertake the ambitious plan his father had suggested – making a complete collection and publishing a description of the island's flora – his mind was changed when he met a local botanical enthusiast called Ronald Campbell Gunn, who had been corresponding with Sir William Hooker for some years.

Born in Cape Town to a soldier family, Gunn had worked for a time as an engineering clerk in Antigua before migrating to Hobart in 1830 to join his brother. Socially ambitious and an able administrator, he'd worked his way from convict superintendent to the position of Governor Franklin's secretary. Gunn joined Sir William Hooker's circle of colonial correspondents in 1832, inspired by the example of a landowner friend who'd been sending seeds and dried plants from his estates. Gunn was attracted to botany both by curiosity and the desire to cultivate a gentlemanly pursuit. In offering to send samples himself, he admitted he knew little about botany and asked, in effect, if Sir William would reciprocate by supplying him with books explaining the complex 'natural system' of identifying plants. Sir William – free of the condescension of his English colleagues, who'd not even bothered to reply to Gunn – did so, and initiated a regular exchange of letters.[6]

By the time Joseph Hooker met Ronald Gunn in August 1840, the self-taught colonial botanist had come a long way. As a police magistrate in 1834–35, he'd patrolled a hundred-mile jurisdiction of half-explored country around Launceston, where he collected scores of rare or unknown species.[7] On his return to Hobart in 1836, these laid the foundations of a private herbarium. He'd also built up an impressive circle of local botanical enthusiasts, which Lady and Sir John Franklin formed into the Natural History Society. This met regularly at Government House to read, discuss and publish scientific papers, and eventually evolved into Tasmania's Royal Society, the first of its kind outside Britain.[8]

In September 1840, Joseph Hooker read a paper for the circle about a fossil tree trunk he'd discovered at Rose Garland in the Derwent Valley. He'd located it by following in Darwin's earlier footsteps, a debt he would one day reverse with interest. Hooker thought Lady Franklin's enthusiasm for promoting science entirely admirable, but found her overbearing manners difficult to stomach. She was fascinated by this 'boyish' young botanist and bombarded him with invitations to dinners, picnics and sightseeing expeditions.[9] 'Lady Franklin . . . does not understand how I hate dancing attendance at Government House,' Hooker fumed in a letter to his father. On a three-day excursion to the convict prison of Port Arthur, he even risked her displeasure by dodging Sunday Service: 'I thought it excusable,' he confessed to Sir William, 'as being my only chance of gathering *Anopterus glandulosus*.' No wonder Lady Franklin found the boy a shade too serious.[10]

On the other hand, Hooker had taken instantly to Ronald Gunn. The two men formed a warm friendship that was to endure throughout their lifetimes. Gunn was generous in sharing his botanical knowledge, expert at preserving dried specimens, and willing to display the right kind of colonial diffidence to the young metropolitan expert. He deferred to Hooker's strong opinions on the need to avoid making hair-splitting distinctions among local plant varieties. As someone interested in making global comparisons of plant distribution, Hooker favoured taxonomic 'lumping' into large groupings, rather than 'splitting' into minutely differentiated ones.

Indeed, the distinction between these two approaches was to become one of Hooker's obsessions. He believed that colonials tended to complicate comparative botanical studies by seeking to declare every tiny local variation a new species. This proliferation of species was anathema to a worldly botanist keen to explore 'philosophical' issues beyond mere naming and classification.

Hooker was happy, however, to proclaim a tree that Gunn showed him in the centre of the island a new species, naming it *Eucalyptus gunnii*, commonly known as the cider gum.[11] He also shared Gunn's concern about protecting indigenous plants and animals. As early as 1836, Gunn had written to Sir William Hooker of his fear that 'many of our animals and birds will become extinct or nearly so'.[12] Emus were already rare and kangaroos had been slaughtered 'in tens of thousands'. Gunn made no mention of the island's marsupial carnivores – the thylacine, today extinct, or the Tasmanian devil, today at risk – even though his own portrait, with its twin peaks of dark spiky hair, rather resembled the latter.

Hooker could recall 'no happier weeks' than those spent wandering in the bush and clambering up mountains with Gunn to collect plants to dissect in his friend's library.[13] Like most of the ship's crew, he began to think of Hobart as a home away from home. When, after three months of re-victualling, the expedition departed Hobart on 12 December 1840, even introverted Joseph Hooker felt himself carried away by the locals' patriotic desire for the expedition to 'eclipse all other nations'.[14]

The 'wonders' of Antarctica began in earnest when they reached the expanse of tessellated ice known as 'the pack' in early January 1841. Huge interlocking slabs of ice barred their path southward as far as the eye could see. Unfazed, Ross began the daunting task of using a sailing ship as an icebreaker. Time and again he pitched the *Erebus*'s reinforced prow against the same spot on the ice crust until it cracked open, allowing the ships to thread their way slowly through a web of half-channels. A century later, the conqueror of the South Pole, Roald Amundsen, would hail the crew of the *Erebus* and *Terror* as pioneers – the first sailors with ships and hearts strong enough to brave the minefield of the pack.[15]

Hooker, oblivious to his future fame, took the opportunity to sketch 'flocks of penguins standing erect with outspread flippers and croaking dismally at our intrusion as they were thrown out of their perpendicular by our smashing against their icy walks'.[16]

After two hundred miles of hammering, warping and sawing through the ice, they found themselves in open sea, and at last Ross was able to set a direct course towards what he believed would be an ocean-based south magnetic pole. His hope proved short-lived. Through thick fog on 11 January, they sighted a distant outline of mountains. Surgeon McCormick excitedly declared it to be 'a new land . . . a Southern Continent'.[17] Somewhere within it lay the south magnetic pole. The snow-capped peaks rose to an estimated height of fourteen thousand feet, but long fangs of ice, a boiling surf and a lee shore prevented any hope of landing. By way of consolation, Ross reported at noon that they had reached latitude 70° 14'S – the first sailors to eclipse James Cook's remarkable southward record of 1774.

The following day, Ross decided to land with a small group of sailors on a rocky island off the coast, so as to take possession of this unreachable territory under the name of 'Victoria Land'. In full ceremonial dress, he gabbled imperial formalities while the sailors huddled around a makeshift flagpole, having squelched through penguin guano so thick it felt like peat. McCormick described the smell as 'stifling'. The blacksmith, Sullivan, claimed that the annexation took place 'without opposition', but this was not quite true.[18] Hordes of penguin elders pecked at the sailors' legs, while younger spectators looked on disapprovingly from tiers of burrows rising three hundred feet up the lava rock-face.

McCormick claimed that the penguin ranks were so 'thickly formed . . . [that] without kicking them to right and left there was no getting through their dense legions'. The penguins were not impressed, issuing 'such harsh notes of defiance . . . that we could scarcely hear each other speak so as to be understood. These sturdy, bold birds, standing erect on their tails, with the horny feathers of both head and neck ruffled in anger, their flipper-like wings extended from their sides, looked altogether the most ludicrous and grotesque objects imaginable.'[19] Such is the ingratitude of natives.

Island landings also presented other hazards. Two weeks later, another attempt to visit a volcanic rock off this coast nearly proved fatal for Hooker. As he leapt ashore from the tossing pinnace, he slipped on the wet, ice-covered rock and was swept into the sea. A lucky grab from one of his shipmates prevented him from being pulped between the boat and the rocks. Chilled, exhausted and shaken, he was sent back to the *Erebus*. Ross returned later to blast the young botanist's hopes of finding even a stray lichen or seaweed in these climes. He'd seen 'not the smallest trace of vegetation' and concluded that 'from the total absence of it, the vegetable kingdom has no representative in Antarctic lands'.[20]

Hooker found some compensation in capturing the new landscapes with pencil and watercolours. The tonal contrasts of paint seemed more apt than words for evoking the strangeness of the world he was encountering. Though he could not match Darwin's abilities as a wordsmith, his sense of composition and visual precision were invaluable assets for assimilating and interpreting unfamiliar vistas.

The afternoon following his misadventure, he was rewarded by 'a spectacle the most inspiring that could be imagined.'[21] At longitude 76° 57'S, all hands crowded on deck to look at two volcanoes looming twelve thousand feet above the ice. The active one, spewing dark smoke and molten larva, they named Mt Erebus, its extinct twin Mt Terror. In later descriptions, Hooker struggled to convey the eerie sublimity of the scene:

> Those living flames . . . that were unceasingly poured out with dense columns of the blackest smoke, we regarded with an emotion that no language could express: for they were not natural

> accompaniments of a region where the sun itself produces no melting of the ice . . . the water and the sky were both as blue, or rather more intensely blue, than I had ever seen them in the tropics, and all the coast was one sparkling mass of which, when the sun approached the horizon, reflected the most brilliant tints of golden yellow and scarlet. Then to see the dark cloud of smoke tinged with broad sheets of flame rising in an unbroken column from the volcano, one side jet black, the other giving off various coloured fire, and sometimes turning to seaward and stretching many miles along the strata of the upper air. This was a sight so surpassing everything that can be conceived and so heightened by the consciousness that we had penetrated to regions far beyond what had been deemed practicable before, that it caused a feeling of awe to steal over us at the contemplation of our comparatively utter insignificance and helplessness . . .[22]

Neither prose nor paint could adequately grasp the next wonder. An intense blink of light on the horizon indicated something novel. It proved to be an ice shelf the size of France, three hundred thousand square miles rising out of the water at a height of a thousand feet, with a far vaster underwater penumbra. It dwarfed their ships, filled their line of vision and blocked their passage to the pole.

Ross declared that they might as well try to sail through the cliffs of Dover, and he called it what it was – the Ice Barrier – though it now carries the blander name of the Ross Ice Shelf. Sailors Sullivan and Savage echoed the opinion of them all: 'It is Quite certain and out of Doubt that from the 78 degree to pole must be one solid continent of Ice and Snow'.[23]

Hooker sent his father a more scientific assessment of its origin and composition:

> The Barrier is probably only a large solid pack filling up a broad shallow bight, like that of Benin or South Australia. Some unusual severe winter, ages ago, first filled it with a sheet of

> ice, and as the snow fell it sunk deeper and deeper every year until it stranded; the sun has no power on it now, and so every snow shower must add to its height. What atmospheric revolutions of centuries may produce we cannot know; but while the climate of the South is so equable and the removal of ice by drifting . . . proportional to its slow drifting accumulation to the South of the Packs, these vast phenomena must remain comparatively unchanged.[24]

He did not mention what would happen if the climate there were ever to become several degrees warmer.

Ross attributed the final marvel of the expedition directly to 'Providence'. On 7 March, the wind dropped to nothing and the heavy easterly swell began to sidle the two ships towards a constellation of eighty-four giant icebergs. After eight hours of drifting, they faced 'what to human eyes appeared inevitable human destruction'.[25] As the gap closed to less than half a mile, 'the roar of the surf, which extended each way as far as we could see, and the crashing of the ice, fell upon our ear with fearful distinctness'.[26] Just then, in answer to terrified prayers – as Ross later represented it – a breeze began to fill their sails and inch them towards open water.

That evening, the Aurora Australis flared in the sky, flinging glorious arcs of light on the bergs that had so recently threatened them. The brilliant pink, green and gold emanations escorted the ships most of the way back to Van Diemen's Land.

For the next three months in Hobart, during the autumn of 1841, while the *Erebus* and *Terror* re-victualled and the crews prepared themselves for a second assault on the ice, Joseph Hooker seized every moment to resume his collecting with Ronald Gunn. At the end of this time, Hooker felt confident that they had gathered a broad representation of the island's flora. Perhaps the possibility of publishing a book describing all the species of Van Diemen's Land might be achievable after all?

Hooker had become so engrossed with collecting that he half considered Gunn's suggestion of feigning illness and staying behind when the expedition prepared to make its second foray into Antarctica.[27] Gunn urged him to take advantage of the opportunity:

> Two years spent in these Colonies would establish your future fame, and the knowledge you would acquire of our peculiar vegetation would be invaluable to you wherever you went afterwards . . . Your expenses here (living with me during the winter and occasionally in summer when not out collecting) would be very trifling and I would get you one or two assigned servants to ramble with you. An old packhorse would complete your establishment and as I know everybody I could furnish you with letters everywhere. At all events and whatever comes to pass, look upon me as your devoted friend, and call upon me to assist you in every way that you think I can be useful, without fearing that I shall back out.[28]

In the end, however, Hooker's sense of duty overcame this strongest of botanical temptations, and he remained with the expedition when it left Hobart for Sydney in July 1841 to make magnetic observations before for their second Antarctic trip.

He realised that the shortness of their stay, along with the work of earlier botanists, would restrict what he could achieve in Sydney. Even so, he once again benefited from his father's colonial links. One of Sir William's naturalist friends, Alexander Macleay, had migrated to New South Wales as Colonial Secretary in 1826, to be followed in 1839 by his eldest son, William Sharp Macleay, an even more distinguished entomologist.

Joseph Hooker didn't think much of Sydney, particularly after he nearly drowned himself by walking off a wharf in the dark.

But he was dazzled by the Macleays' grand house overlooking Elizabeth Bay. It seemed to him 'a botanist's paradise': the workshop was crammed with local specimens, the grounds tasteful, the garden filled with exotic plants and parrots. A sea eagle even soared operatically over the waters of the harbour.[29] William Sharp Macleay topped Hooker's pleasure by promising to send him mosses and other lesser-known mainland Australian plants.

From Sydney they sailed to their last pre-polar destination, the Bay of Islands on the North Island of New Zealand, arriving on 17 August 1841. Moving slowly past the whaling port of Kororareka in a welter of fog, wind and rain, they anchored at the mouth of the Kawa Kawa River, not far from Murderers Bay, where the French Captain Marion du Fresne had been killed and eaten by Maori warriors in 1772.[30] Yet on reaching this beautiful bay, Joseph Hooker felt as depressed as Darwin had six years earlier. His problem, though, was not homesickness, but scientific frustration.

The Ross expedition's original plan – to winter in the Antarctic Circle, then collect plants in the spring when the snow melted – had been shelved by the unexpected harshness of the southern icecap. It had proved to be a world without plants, a desert of ice where not even reinforced ships could hope to survive the jaws of the midwinter freeze. Hooker was afraid that their forthcoming polar expedition would be as botanically bleak as the first: he'd again have to spend all his time sketching icescapes and working on maritime zoology – much to his father's displeasure.

Even beyond these forebodings, Hooker was suffering a crisis of confidence. Sir William's letters had stressed the importance of collecting in the nearly virgin countryside of New Zealand, but Hooker felt overwhelmed by the scale of the task. At the end of the first week, he wrote a letter to his father berating himself for his lack of opportunities to work on botany. 'Could I with honour leave the expedition here,' he wrote miserably, 'I would at once and send home my plants for sale . . . but now my earnest wish is to be able on my return home to devote my time solely to Botany . . .'[31]

As in Van Diemen's Land, Joseph was rescued by a colonial. Having presented a letter of introduction to a local plant enthusiast, William Colenso, Hooker realised he'd found a New Zealand equivalent of Ronald Gunn. True, they were not exactly alike. Tall, morally intense and often abrasive, Colenso had large intellectual ambitions and little deference. He was an evangelically devout Cornishman who'd migrated to nearby Paihia in the early 1830s to work as a printer for the British and Foreign Bible Society. Liking both the countryside and the indigenous people, he'd translated and printed Maori-language religious texts as part of a larger drive to become a missionary. An encounter with Charles Darwin in 1835 had awakened his interest in natural history, which was further encouraged by befriending a prominent New South Wales botanist, Alan Cunningham.

On joining Sir William's circle of correspondents in the mid-1830s, Colenso had proved himself as prickly as he was useful. He refused to give up the presumption that his intimate knowledge of living New Zealand plants gave him as much right to name local species as London botanists who made their identifications from dried specimens.[32] He also urged recognition of Maori botanical expertise, something most British botanists thought ridiculous.

What mattered to Joseph Hooker, however, was that Colenso volunteered to help him gather and identify New Zealand flora. The man had an impressive dried plant collection, a useful garden of local plants, a boat for navigating rivers, and he was prepared to act as a guide and translator of the local Maori. In short, Colenso was Joseph Hooker's passport to the botany of New Zealand.

Over the next three months, Hooker, Colenso, and a variety of shipmate helpers, including the assistant surgeon of the *Terror,* David Lyall, explored the beaches, rivers, streams, swamps, lakes, hot springs, waterfalls and mountains around the Bay of Islands. As in Van Diemen's Land, Hooker felt confident about specialising in non-flowering plants, because of Colenso's mastery of flowering species. The latter's geographical knowledge helped Hooker find rich botanical pickings on the banks of the Waitangi, Waikare, Waimate and Kawa Kawa rivers.

On one occasion, he was able to collect fourteen species of lichen from a single tree; on another, he gathered ten species of fern within a radius of a few hundred yards.[33] They tramped through the dense kauri forests between Waitangi and Waimate, and, like Darwin earlier, measured the girth and height of the beautiful, smooth-trunked trees that rose sixty feet before throwing out horizontal limbs.[34] Unlike Darwin, though, Hooker loved the landscape and was respectfully impressed by Maori culture. With Colenso's help, he discovered the native names of a number of botanical species, and he investigated the Maoris' use of several local flax species, out of which they manufactured fabrics and mats. The plant looked to have definite economic promise.[35]

Colenso was also able to broker a good relationship with the local Maori. Since the signing of the Treaty of Waitangi the previous year, tensions had built up in the district because prominent Maori believed they'd been tricked into ceding their lands. Colenso had tried, albeit unsuccessfully, to bring their concerns to the attention of the government, which is probably why he and Hooker experienced no trouble during their collecting expeditions.

Hooker did get a taste of how difficult things might have been without his mediator. In late November, a few hours after he'd been collecting at Paihia Bay, Maori warriors burnt down the nearby house of a whaler's widow, killing her and one child and abducting another. The warriors believed that with her death the land would revert to its traditional owners. News of the murder sent rumours of insurrection through the frontier port of Kororareka, and Captain

Ross received an urgent delegation requesting protection.[36] With time pressing, he had to refuse, and the expedition set sail for Antarctica a few days later, on 23 November 1841.

Despite this sobering conclusion to their three-month stay in New Zealand, Joseph Hooker warily allowed himself to think it a success. Shortly before they left, he was excited to learn from the papers that his father had been appointed Director of Kew Gardens. Though still poorly paid, Sir William now worked at the hub of British botany. Ross, in celebration, bent the Admiralty rules and gave permission for Hooker to send a large parcel of New Zealand specimens directly to Kew – unaware that he had been secretly doing so already. Officially, all specimens were supposed to go to the Admiralty. With some trepidation, Hooker suggested in an accompanying letter that he might have accumulated enough specimens to publish a flora of this fascinating country. Thankfully he was right – as Charles Darwin would one day deeply appreciate.

In late December 1840, not long after departing the Auckland and Campbell islands to the south of New Zealand, the *Erebus* and *Terror* reached the pack ice. Within a week they found themselves trapped. Given the time of year, Captain Ross decided to make the most of the occasion, and so on New Year's Eve 1841 a few white terns and a handful of Antarctic penguins witnessed an odd event.

Two hundred and fifty miles within the pack ice, at latitude 66° 32'S and longitude 156° 28'W, the two shiploads of Britons held an ice party. First the two ships were lashed together with hawsers and secured with double anchors to a large iceberg, then a contingent walked out onto the ice to prepare the site for a grand ball and circus.

Even in this relaxed atmosphere, the tasks were divided according to naval rank. As the expedition's most accomplished artists, assistant surgeon Joseph Hooker and his good friend J.E. Davis, second master of the *Terror*, ornamented an ice ballroom by sculpting an eight-foot seated Venus de Medici. After this they turned their attention to moulding sofas, tables, goblets and grog glasses. The most ambitious ice architecture, however, was conceded to the sailors under the charge of the boatswain. They built a tavern, complete with a taproom, coffee room and an entertainment arena. In an inclusive spirit, the ice pub carried a sign intended to entice both officers and men: one side read *Pioneers of Science*, the other *The Pilgrim of the Ocean*. On a flagpole nearby was hoisted the silk Union Jack that Ross had flown when he discovered the north magnetic pole a decade earlier.

After dinner, Captain Crozier of the *Terror*, in full naval uniform and with tall, dark-haired 'Miss Ross' of the *Erebus* on his arm, entered the tavern to a welcoming fusillade of rifle shots. The two spliced the mainbrace with the men – sailor's cant for downing quantities of grog – then the ball began. The egregious couple opened proceedings with a stately quadrille, which was followed by wild reels and country dances. 'You would have laughed,' Davis later wrote to his sister, 'to have seen the whole of us with thick overhaul boots dancing, waltzing and slipping about . . . the ladies fainting with cigars in their mouths, to cure which the gentlemen politely thrust a piece of ice down her back . . .'[37]

As the ball degenerated into rowdy games, the sailors took over. The joint account of Cornelius Sullivan and James Savage, respectively blacksmith and able seaman on the *Erebus*, boasted that 'the Exhibition in the Circus by far exceeded the waltzing in the ballroom . . . When the essence of barley heated our Gents the snow balls went flying. After a round of coffee they [the officers] withdrew from this rare scene of mirth. So that the Tavern Tap and ballroom's half empty bottles and in fact the whole ice berg belonged to the Jolly Tars until morning.'[38] Ross's subsequent claim that 'we were a very happy party' seemed something of an understatement.[39]

On 19 January 1842, they experienced their first hurricane in the pack ice. For nearly two days, recorded Ross, 'our ships were involved in an ocean of rolling fragments of ice, hard as floating rocks of granite, which were dashed against them by the waves with so much violence that their masts quivered as if they would fall with every successive blow; and the destruction of the ships seemed inevitable'.[40] The ice missiles stripped the copper lining off their hull 'like brown paper', and Hooker, though always cool in a crisis, genuinely feared that the *Erebus* would be ground to powder.[41]

When the wind eventually died down, both ships – now carrying broken rudders, splintered spars and torn sails – found themselves back where they had first entered the pack. The fundamental strength of their hulls and the ingenuity of their carpenters enabled the two captains to begin again, but it was not until 2 February that they finally broke through into open water. They had been stuck in the jaws of the ice for fifty-six days.[42]

Another twenty days of stormy sailing saw them tacking eastward alongside the Ice Barrier, in the hope that they could edge closer to the pole. This they did, marginally. At latitude 78° 11'S and longitude 161° 27'W, they just pipped their previous southward record, before the Ice Barrier began to push them northward again. With small fish freezing solid in their bow-wave, an exhausted Ross decided to head towards their winter quarters in the Falklands Islands.

On 13 March, just when they glimpsed clear water through a haze of ice and snow, a massive iceberg materialised in front of the *Erebus*. Ross swung the ship abruptly to the wind in an effort to avoid the monster. At the same moment, they saw the *Terror* running down on them at speed under heavy sail. As Ross recorded:

> [T]he concussion when she struck us was such as to throw almost everyone off his feet; our bowsprit, fore topmast, and other smaller spars were carried away; and the ships hanging together, entangled by their rigging, and dashing against each other with fearful violence, were falling upon the weather face of the lofty berg under our lee, against which the waves were breaking and foaming to near the summit of the perpendicular cliffs. Sometimes she rose high above us, almost exposing her keel to view, and again descended as we in our turn rose to the top of the wave, threatening to bury her beneath us, whilst the crashing of the breaking upper-works and boats increased the horror of the scene.[43]

Having hacked clear of the *Terror*, Ross and his crew faced a breathtaking risk. Despite the gale-force winds, the captain decided to execute a 'sternboard'. In effect, he released the sails, reversed the helm and sailed the ship backwards, 'plunging her stern into the sea'.[44] With another iceberg sweeping down on them, they managed to bring the disabled ship's head around to steer between the two bergs, through a gap about twice the ship's width. Ross stood stock-still on the quarterdeck, his face white and drawn and his arms folded, watching their yards scrape on the sides of the ice cliffs, until 'the next moment we were in smooth water'.

The *Terror*, meanwhile, was navigating through an identical gap. '[N]ever till that moment did I in reality know what fear was,' the master's mate J. E. Davis later wrote to his sister. 'Oh it was horrible, truly horrible. That time will never be effaced from my memory.'[45] One man went out of his mind with fear and several others began crying like babies, but most remained composed. Afterwards, blacksmith Sullivan summed up the general feeling: 'we escaped through Providence one of the most frightful Cases of Shipwreck'd that

Ever occurred on the high seas . . . we poor Pilgrims of the Ocean thought it was our Last in this life'.[46]

Despite their joy at surviving, the collision marked the beginning of general disillusion with the expedition. Up until now, the officers and men had been sustained by their determination to outdo the French and Americans. Both expeditions into Antarctica had achieved this by sailing further south and discovering more land than either of their rivals, but it now seemed there was nothing more to prove – only lives to be lost. Moreover, instead of being allowed to recover their health and spirits in vibrant Rio, Ross insisted that they winter in the miserable Falklands – a decision, Hooker thought, based on a fear of wholesale desertions.[47]

The Falkland Islands, with their population of around sixty families, offered no temptation to jump ship. There was nothing for the sailors to do in this boggy, windswept place except get drunk on rum. Fresh vegetables were scarce and meat could only be had if you were prepared to hunt the wild cattle – a perilous and grisly process, as Hooker described it. Ross, who was still a mass of nerves after their recent ordeal, was inclined to fly into unprovoked rages, which did not endear him to the Governor of the Falklands, young Lieutenant Moody. The two quarrelled violently and Hooker found himself caught uneasily in between.

Most of the ships' officers grew equally morose, especially when news reached them that the Admiralty had promoted only a few of their number. Hooker, though personally uninvolved, wrote his father 'a hearty growl . . . for the sake of my shipmates'. They had done their duty under the harshest conditions with no thanks from the Admiralty. Ross even refused to allow them to communicate any of the expedition's successes to the outside world. The general view was that he wanted all the glory for himself. As a result of this selfishness, Hooker grumbled, the world had forgotten they existed. 'Naturally,' he told Sir William, the crew 'look very coldly' on the forthcoming third cruise to tackle Antarctica once again.[48] 'Jimmy', as Hooker and his officer friends now impertinently referred to Ross in private, was proving himself to be puffed up with vanity.

Still, Joseph Hooker enjoyed both their stopover at the Falklands and a subsequent spell at Tierra del Fuego, where they called before again heading south. On disembarking at Port Louis in the Falklands on 6 April 1842, he'd been relieved at last to receive Sir William's congratulatory letters on the excellence of his sketches, notes and plant collections from Kerguelen's Land and Van Diemen's Land. Perhaps he might have the makings of a sound botanist after all?

Hooker could also offer his father a new gift from the Falklands: an intriguing and potentially profitable variety of tussock grass. When fertilised by penguins, elephant seals and seaweeds, the tussocks grew to resemble miniature palm trees around six feet high. Local cattle and horses thrived on their thatch-like fronds, and Governor Moody had high hopes the grass could become a source of feed for domestic animals in Scotland and Ireland. Hooker's botanical description later helped Moody win a gold medal from the Royal Agricultural Society, and greatly excited Sir William at Kew Gardens. Farmers from all over Britain began badgering the Gardens for seeds.

These two winter stopovers also provided Hooker with the data he needed to develop plausible theories about the origins and distribution of plant species on the islands of Antarctica, something he and Darwin would debate in the future. During their first push into Antarctica, Hooker had collected plants at the Auckland Islands, more than two hundred miles south-west of New Zealand, and at Campbell Island, further south still. By combining all this material with his new observations in the Falklands, and on Hermite Island, just off the tip of Cape Horn, he began to detect some patterns. At Hermite Island, he found plants that were, he believed, 'the hardiest of their race in the southern hemisphere'. The landscape here was exactly like the west of Scotland: one had only to clothe

the scenery of his boyhood with the analogous plants of this island to 'understand the relations, in habit and station, which . . . these bear to one another'.[49]

The islands near Tierra del Fuego also surprised him in a more fundamental way. He discovered that all the Antarctic islands he'd explored up to this point 'have borrowed many plants from this, the great botanical centre of the Antarctic Ocean', even though some were more than five thousand miles away. A number of English plants had somehow managed the long migration, 'though 106 degrees of ocean roll between, and some of the species in question inhabit no intermediate latitudes'.[50] His imagination took fire at the idea 'that plants, found in these isolated spots alone must have traversed . . . thousands of miles of the stormiest ocean of our globe'. Such observations, he ruminated, 'perpetually draw the traveller's mind to that interesting subject – the diffusion of species over the surface of the earth'.[51]

Though he had not yet worked out the key to this oceanic transmission, over which he and Darwin would argue for many years, it was clear that creation theory had no explanation for why so harsh and remote an environment should prove so rich in global species.

Tramping in the Falklands and Tierra del Fuego – the point, as it happened, where he was able to resume his connection with the path of the *Beagle* – Hooker was freshly conscious of how 'all Darwin's remarks are so true and so graphic wherever we go . . . [he] is not only indispensable but a delightful companion and guide'.[52] Towards the end of this long sojourn in the Southern Hemisphere, Hooker found himself generating theories with a fluency that rivalled his hero. Reading Darwin had inspired him in the same way that reading Lyell had inspired Darwin. Hooker could now assemble tables that predicted the way in which the number of grasses, fungi, lichens and mosses would diminish or increase in relation to each other within different climatic zones.

He'd discovered, too, that the distribution and abundance of plants in these regions depended 'not on height of the mean temperature but on the moisture in the air and the equable level of heat and cold free of extremes'. These original observations he dispatched to the treasurer of the Linnean Society, Dr Francis Boott, with instructions not to divulge them. Hooker intended to confound those 'critics who claimed that nothing had been done toward investigating the cause of difference in geographical distribution since the publication of Humboldt's work'.[53]

Such self-confidence was a far cry from the young man who'd so long doubted whether he could live up to either Darwin's achievements or his father's expectations. After five years of research in the Southern Hemisphere, Joseph Hooker now saw himself as having completed the transition from collector to philosophical botanist. A letter written to his father on the eve of his third Antarctic expedition scorned those narrow botanists whose only interest was in tagging their names to new species: 'I should be far more proud of placing a well-known plant in a true position and relation to others,' he declared loftily. There was even a hint of condescension in his suggestion in the same letter that he was happy to help Sir William take up work on the biogeographical distribution of ferns by furnishing him with tables 'in relation to the southern Islands'.[54]

Their third and last expedition into the ice, from January to early March 1843, proved as futile as Hooker and the other officers had feared. They attempted to follow the British whaler Weddell's route south but hit the ubiquitous pack ice, which was this time denser and more extensive than ever before. Everyone privately believed that the voyage was gratuitous. It achieved nothing new, and it inflicted on men already past the point of endurance 'the worst season of the three, one of constant gales, fogs and snowstorms'. Hooker recalled that 'Officers and men slept with their ears open, listening for the look-out man's cry of "Berg ahead" followed by "All hands on deck"'. And the men lay wide awake all night on their chests 'alive to every sound on deck'.[55]

Their nerves were shredded and many drank to compensate. Ross's and Crozier's hands shook uncontrollably. For the remainder of his life, even steely Joseph Hooker could never think of the collision with the *Terror* without feeling physically ill. He reported to his family that Captain Ross 'says he would not conduct another expedition to the South Pole for any money and a pension to boot. Nor would any individual of us join if he did.'[56]

By the beginning of March, violent storms had forced them to retreat. Eleven days later, they crossed the Antarctic Circle for the last time. Hooker gave his father a bleak summation:

> And now we may consider the voyage as over . . . Botany has alone rendered many months tolerable which otherwise would have been to me, as they were others, scarcely endurable . . . As it is I am the only one who can look back on pleasant hours spent on board the *Erebus*, [and] almost the only one who has not 50 times over wailed the day he ever joined the fair fame of the Expedition.[57]

On 4 September 1843, they docked at Folkstone on the south-east coast of Kent. Hooker's family, like the Darwins, found the traveller changed almost beyond recognition. Croaky Joe's arms were muscular from hauling ropes and setting sails, and his chest 'rings almost when struck'.[58] Among the many specimens waiting for Hooker's attention was a small consignment of Tierra del Fuegian plants that had been sent to him by Charles Darwin. He had heard something of Hooker's achievements from Sir William.

A month later, Darwin introduced himself in person to the young botanical enthusiast who so passionately shared his interest in the islands of the Southern Hemisphere.

# PART THREE

# Thomas Huxley and the Voyage of the *Rattlesnake*, 1846–50

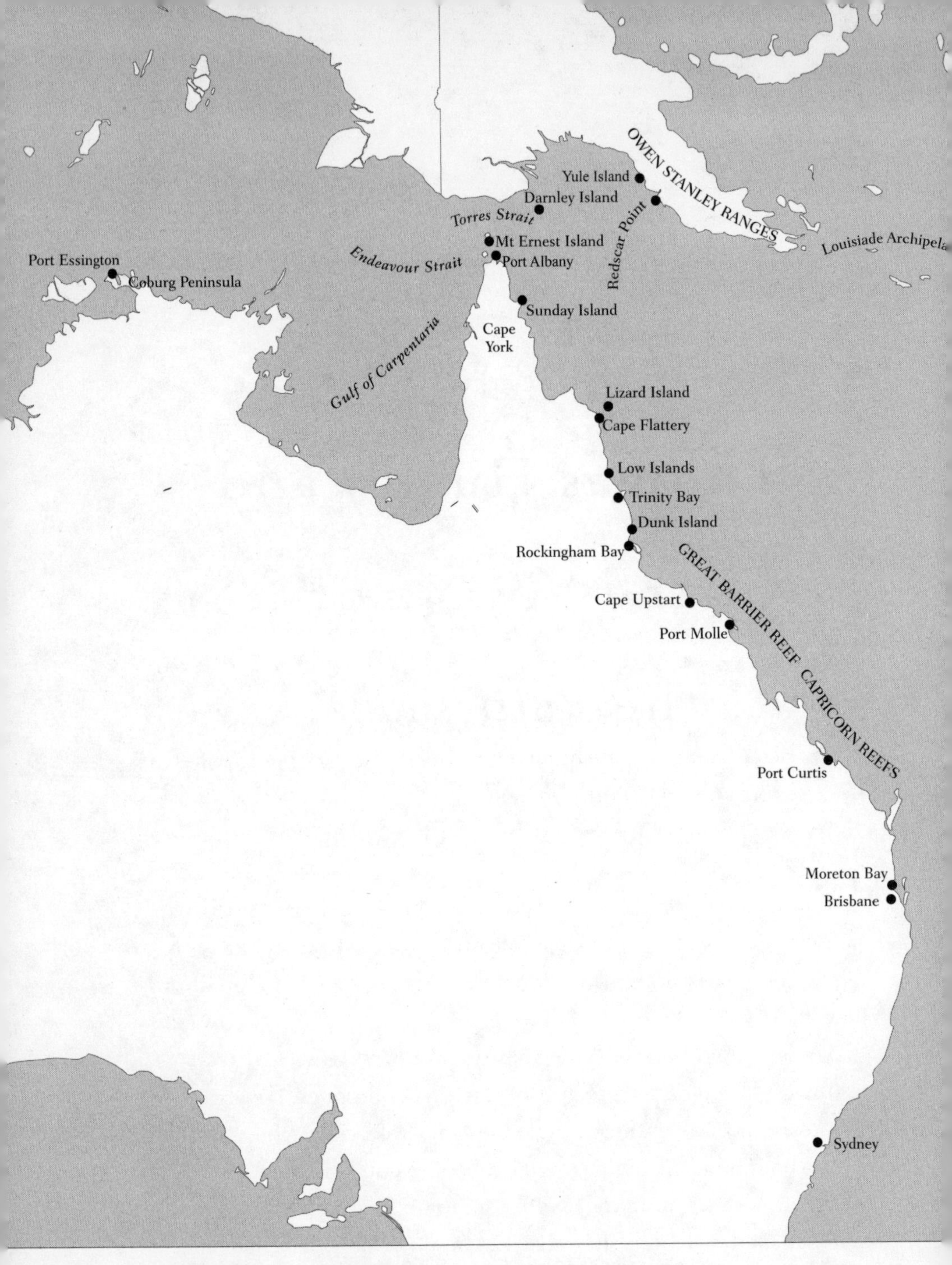

OWEN STANLEY RANGES
Yule Island
Darnley Island
Torres Strait
Redscar Point
Mt Ernest Island
Port Albany
Endeavour Strait
Louisiade Archipela
Port Essington
Coburg Peninsula
Sunday Island
Cape
York
Gulf of Carpentaria
Lizard Island
Cape Flattery
Low Islands
Trinity Bay
Dunk Island
Rockingham Bay
GREAT BARRIER REEF
Cape Upstart
Port Molle
CAPRICORN REEFS
Port Curtis
Moreton Bay
Brisbane
Sydney

THE AREA SURVEYED BY THE *RATTLESNAKE*

# Love and Jellyfish

On 4 May 1847, a lanky, dark-haired man stood on the deck of HMS *Rattlesnake*, a British survey vessel, as it was being towed by a tug around the northern point of the island of Mauritius. Today was Thomas Henry Huxley's twenty-second birthday, and as he watched the landscape drift past he compared the island to the descriptions Charles Darwin had written eleven years before. This little island in the Indian Ocean off Africa was a striking vista, certainly, but it was not, Huxley thought, as lush as the great naturalist had made out.

No one could have guessed that this young surgeon's mate was destined to become the greatest of Darwin's colleagues: his instinctive resistance to authority was more notable. Like the apostle after whom he liked to imagine he'd been named, Thomas Huxley accepted nothing on faith. Every assumption had to be tested and analysed, and was more often than not found wanting. Even his own birth was not exempt. He wrote in his diary that May evening:

> Twenty-two years ago, I entered this world a pulpy mass of capabilities, as yet unknown and save by motherly affection uncared for. And had it not been better altogether had I been crushed and trodden out at once? Nourishing me up was as though one should pick up a stray egg, unconscious whether dove's or serpent's, and carefully incubate it. And here I am what a score of years in the world have made me – such a bundle of glorious and inglorious contradictions as men call a man.[1]

These glorious contradictions included a vein of romanticism that made young Huxley more like Charles Darwin than his cynical self-reflection would suggest. By the time he set foot on shore at Port Louis, he'd come around to conceding the truth of Darwin's opinion of the Isle de France (Mauritius). In a letter to his mother, he called the island 'a complete paradise'.[2] The site chosen for the town was superb. It lay within a natural amphitheatre framed on one side by the spurs and peaks of two great mountains, Peter Botte and La Pouce. At their base lapped the blue waters of the Indian Ocean.[3] Like Darwin, he delighted in the cleanliness of the streets and the varieties of local costume: at the marketplace he saw Parisians in satins, Hindoos in muslin drapery, turbaned Mussulmen, Chinese with pigtails tucked in their caps, and uniformed black African policemen in fezzes.

Even more picturesque was the nearby village of Pamplemousses. Huxley walked there later to visit the island's most romantic monument – the tombs of the two local lovers on whom the writer Bernardin de Saint-Pierre had based his bestselling romance *Paul and Virginia*.[4] Huxley remembered 'the lively feelings . . . excited in me' when, as a boy, he first read the tear-jerker about the girl who drowned off a coral beach and the fiancé who pined to death as a result. He doubted now whether he could recapture those naïve emotions, 'intercourse with the world, the flesh and the devil having extinguished any latent spark of sentimentalism . . . in my bosom'.[5] In retrospect, Paul struck him as a bit of a 'pump' and Virginia as a 'prude'.

Still, Darwin, too, had reread the book 'to procure a small stock of sentiment' before his pilgrimage there in 1836; it was required of every visitor.[6] Despite Huxley's half-suspicion that the tombs were fake, he enjoyed the 'perfect wilderness' of the garden and the charm of two child guides – a latter-day Paul and Virginia – who cautioned him against relic-hunting.[7] The two roses he plucked from the garden did not of course count as relics – and he was not to know that they were destined to become tokens for a love of his own.

Thomas Huxley's shipmates would have been surprised by this 'tender and sentimental' side of his personality. To the *Rattlesnake*'s senior officer, Lieutenant Joseph Dayman, Huxley seemed to be cast in the same austere and fanatic mould as Joseph Hooker, who had served with him on the *Erebus* four years previously. As products of a common training, some similarity between the two young assistant surgeons was to be expected. Eight years younger than Hooker, Huxley had also read boys' stories about the South Seas that filled him with an 'inexpressible longing' to find 'a southern cloud land full of strange wonders and overflowing with adventures'.[8]

Also like Hooker, Huxley had worked manically to pass the first part of a Bachelor of Medicine, earning several prizes on the way, in botany, physiology and anatomy. For all its shortcomings, medicine had been an inevitable choice because both his brothers-in-law were doctors who could offer him partial apprenticeships. He further resembled Hooker in having made a successful application for a position of assistant surgeon at the Haslar Naval School under Dr John Richardson, now Sir John. Discerning the boy's abilities, Richardson recommended him in May 1846 to a survey captain, Owen Stanley, who was cast in a similar scientific mould to FitzRoy and Ross.

Stanley had been commissioned to sail the 28-gun frigate HMS

*Rattlesnake* into the Southern Hemisphere. His task was to survey the Coral Sea around the Louisiade Archipelago of New Guinea, and along the inner side of the Great Barrier Reef. With commercial steamships in the offing, the increasing imperial trade between Australia, China and India made it imperative to discover a safe 'outer passage' between the northern tip of Australia and New Guinea, as well as a reliable inner passage through the more sheltered Barrier Reef.

It was an exciting prospect. Apart from anything else, Huxley told his favourite sister, Lizzie Salt, 'New Guinea . . . is a place almost unknown . . . our object is to bring back a full account of its Geography, Geology and Natural History.'[9]

Huxley's initial pleasure at meeting his future captain echoed that of Darwin and Hooker a few years earlier. Owen Stanley was polite, if a shade aloof, as one would expect from a born aristocrat with a bishop for a father, but he evidently knew his business. Although only thirty-five, the stocky little seaman was worn by years of naval combat, Arctic exploration, and surveying in South America and Australia. He was said, as a result, to have a short fuse, but Huxley noticed only that the captain had been elected a Fellow of the Royal Society and was 'a thorough scientific enthusiast . . . -altogether very much disposed to forward my views in every possible way'. Stanley surprised his new assistant surgeon by promising that research achievements on the voyage would bring promotion. And he'd gone out of his way to demonstrate the sincerity of his support for science, having requested a consignment of reference books from the Admiralty and ordered that the ship's poop be converted into a chartroom suitable for scientific work.[10]

Huxley's own cabin might only be an alcove off the gunroom, four foot, eleven inches high, but he could at least dispose of his 'superfluous foot' of height by sitting with his head sticking out of a skylight.[11]

His first letters home to his sister showed him at his most

cheerful. More typical of his usual attitude to naval authority was the note he sent to his mother complaining that he'd been temporarily accommodated in a rotting, overcrowded hulk while the *Rattlesnake* was being refitted. Appended to the letter was a sketch of himself crouched in a tiny space that resembled the quarters of a slave ship. Underneath he'd scribbled the famous abolitionist motto: 'Am I not a man and a brother?'[12]

As this suggests, Thomas Huxley was predisposed to mordant satire – one of several ways in which he differed from his two future friends Joseph Hooker and Charles Darwin. This sardonic strain was more than a mere accident of temperament: his flinty personality had been shaped by twenty-two years of struggle.

If Darwin was born into the wealthy gentry and Hooker into the respectable professional class, Huxley was a product of what his English contemporary Edward Gibbon Wakefield called the 'uneasy middling classes'.[13] As an advocate of assisted immigration to the colonies, Wakefield was thoroughly familiar with the shabby clerks, teachers, shopkeepers and artisans who battled to maintain an independent living during the most volatile period of Britain's industrial, agricultural and commercial revolutions. Thomas Huxley had been born in 1825 over a butcher's shop outside London, in Ealing, where his father worked as a mathematics teacher at an ailing Church of England private school. Thomas inherited George Huxley's drawing talent, short temper and obstinacy, but nothing else.

By the time Thomas attended his father's school at the age of eight, it was in terminal decline. The boy's only two years of formal schooling ended when George Huxley was forced to move to Coventry to work as manager of a small, shaky savings bank. As the youngest surviving child of eight, Thomas clung desperately to his mother during this dislocation. Rachel Huxley had endowed him with glittering black eyes, a fierce moral intensity and a quick intelligence, but she could give him little attention.[14]

Reverend Thomas Malthus's well-known thesis – that overpopulation would inevitably press on resources to produce poverty – was perfectly illustrated by families like the Huxleys. Thomas grew up poor, neglected and sad, as remote from his brothers as if they were strangers. Love kindled only with his oldest sister Lizzie, who became a substitute mother.

The clever, oversensitive boy turned in on himself, becoming, as he said, 'one of the most secretive, thin-skinned mortals in the world', who 'had little pleasure in the general pursuits of boys of my own age'. He decided early that 'joys and anticipations derived from the inward world . . . and sorrows and misfortunes . . . from the outward world'.[15] His sarcastic expression, brooding eyes and beaklike nose made him look raptorial, which is exactly how most people found him. His childhood pleasures came from performing homemade scientific experiments and reading. Without guidance, he tackled the most challenging and subversive of adult books. James Hutton's *Theory of Earth* introduced him, at the age of twelve, to the immensity of geological time. Sir William Hamilton's *Logic* triggered an appetite for inductive method and metaphysical theory. Thomas Carlyle's thundering secular prophecies inspired him most. The crabby historian argued that leaders with passion and genius were needed to purge the mechanical spirit that crushed and dehumanised the British poor.

Young Huxley decided to become such a Carlylean hero, confessing many years later to the social reformer Beatrice Webb that 'though he had no definite purpose in life he felt power; [and] was convinced . . . he would be a leader'.[16] He even learnt German so as to share Carlyle's passion for German Romantic literature.

Like so many of England's uneasy middling classes, Huxley found himself drawn to liberal ideas and values. All around him in the fast-industrialising Midlands town of Coventry were textile workers who'd embraced the mass democratic program of the People's

Charter, a program of political reform that included universal suffrage and annual parliaments. To many of Britain's landed elite, this was nothing less than an incitement to revolution.

Such iconoclasm extended also to religion. In plain brick chapels, dissenting preachers from Protestant denominations outside the established Anglican Church ranted against the corruptions of 'political priestcraft'. These inheritors of the truculent independence of the Puritans were angered that Anglican clergymen served as magistrates to dole out capital punishment to the poor, or as bishops to vote in the House of Lords against all reform measures. Small capitalist silk- and ribbon-makers from these same dissenting denominations – Baptists, Congregationalists, Independents, Unitarians, and many others – also agitated to end the longstanding Anglican control of local councils within the booming industrial towns.[17]

In such a political climate, Huxley learned to subject the claims of scripture to the forensic test of reason, and to sympathise with the plight of working men and women. His hatred of unmerited social privilege grew almost visceral, especially when this privilege took the form of a clerical monopoly over the few positions available in the scientific world. At Oxford and Cambridge, every professor had to pass a religious test based on the articles of the Anglican Church. Whether or not Huxley called himself a 'Chartist' or a 'radical reformer', he shared the radical's typical desire to take 'the axe to the root' of all forms of clerical, aristocratic and monarchical corruption. He wanted to see a merit-based world where the 'useful', productive and labouring classes received their proper worth.[18]

Struggles against privilege reached into Huxley's own occupation of medicine. The College of Surgeons operated as an exclusive gentlemen's club while lesser surgeons, general practitioners and drug-making apothecaries worked like lowly tradesmen. Not a few apothecaries also enrolled in radical ranks.

Unable to become a mechanical engineer as he had wanted, Thomas Huxley joined an older brother, James, as an apprentice to his brothers-in-law John Salt and John Cooke. Both practised within the scruffy fringes of London medicine. Here Huxley's life as a 'hospital walker' was a world away from the student experiences of Darwin and Hooker at Edinburgh and Glasgow University. He attended lectures at a poor man's cramming institution in London, slaved at his books to win a Charing Cross Hospital scholarship for the sons of impecunious gentlemen, and dispensed drugs to the outcast poor in the dockyard slums of Rotherhithe on the River Thames.

Every day, Huxley walked to work at the tiny dockyard surgery through 'alleys nine or ten feet wide . . . with tall houses full of squalid drunken men and women, and the pavement strewed with still more squalid children. The place of air was taken by a steam of filthy exhalations . . .' He treated starving and angry women piece-workers from the clothing trade, from whom he learnt that only hunger and disease prevented the downtrodden from rising up to plunder the rich.[19]

Somehow, through all this, he managed to pass his Bachelor of Medicine with distinction. He also found in himself an unsuspected talent as a researcher. Working one day with his favourite lecturer in physiology, Thomas Wharton Jones, young Huxley discovered a tiny new membrane attached to the hair follicles of the human head. It became known as Huxley's Layer.[20] He had been bitten by the scientist's bug.

Huxley lacked influential friends and relatives to foster his path into the navy, which, as it had been for Hooker, was the only way he could hope to earn money as a doctor and pursue scientific research. Naturalism might not be particularly important to the Admiralty, but at least it was permitted as a sort of ancillary hobby. So on 31 January 1846, he wrote bluntly to the Physician-General of the Navy, Sir William Burnett: 'Having a great desire to enter the Medical Department of Her Majesty's Naval Service and being at the same time totally unprovided with any friendly influence by which the attainment of my object might be accelerated – I take the liberty of addressing myself directly to you as the Head of the Department.'[21]

When they met in interview, Huxley stressed that he needed the job as much for the pay – a respectable seven shillings and sixpence per day – as for the chance to make a mark in science. He was deep in debt to members of his family, from whom he'd been borrowing two pounds a week for three years to cover food and rent. This air of desperation led Burnett to ask whether he was Irish – one social handicap Huxley could disclaim.

Huxley was fortunate that the breath of merit-based administrative reform had reached the navy's medical department, even though it co-existed with the traditional norms of influence that had benefited both Darwin and Hooker. Because young Huxley's academic record was impressive and he passed the naval entrance exam easily, Burnett and Richardson decided to admit him to the service. On 13 May 1846, Thomas Huxley became a surgeon's mate, free to spend his first pay cheque on a microscope, books, and a fancy naval uniform – an action that only compounded his family debts.[22]

Always self-aware, Huxley realised that his limited education must shape the kind of science he could do on the voyage. His scientific training was narrow: physiology, 'the mechanical engineering of living machines', had been the only part of medicine that truly interested him. 'I am afraid,' he later admitted, 'there is very little of the genuine naturalist in me. I have never collected anything, and species work was always a burden to me: what I cared for was the architectural and engineering side of the business, the working out of the wonderful unity of plan in the thousands and thousands of diverse living constructions, and the modifications of similar apparatuses to serve diverse ends.'[23]

There was no question of using the voyage to publish a naturalist travel journal like Darwin's; neither would he waste time trying to discover new species or build specialised collections as Hooker had. Thomas Huxley had to itch where he could scratch. He determined to analyse the comparative structures, or morphologies, of living species that had been 'long known but very little understood'.[24]

Finding the biological links between rare marine creatures was an ambition worth pursuing. It would make use of his engineering talents, interest famous anatomists like Professor Richard Owen, and be a feasible form of study on a small, crowded ship. He would get advice on which marine species to study by asking Professor Owen and other experts.[25]

On 10 December 1846, while the *Rattlesnake* was still taking last-minute supplies at Plymouth Sound, Huxley thus opened his diary with 'the plan I must pursue so as to turn to the utmost advantage the opportunities this expedition will afford me . . . Not only my previous habits and tastes but the nature of the accommodation and opportunities afforded by this ship clearly point to the habits and structure of the more perishable or rare marine productions as the most likely to be profitable.' University experts with large libraries had all the advantages for identifying and naming such creatures, but 'what I *can* do and they *cannot* is: I can observe 1. the "habits" of living bodies, 2. their mode of development and generation, 3. their anatomy by dissection of fresh specimens, 4. their histology [living tissues and organs] by microscopic observation.' After this, Huxley listed the genera of pelagic (open sea) marine creatures whose nervous systems, structures and modes of generation he would lay bare, a kaleidoscope of molluscs, jellyfish, stingers, squirts, barnacles, sea slugs, sea snails, worms, anemones, starfish and corals.[26]

It was the London biologist Edward Forbes who persuaded the young surgeon's mate that studying jellyfish and other perishable marine creatures would offer him the best chance of making a name for himself. Huxley took an instant liking to this genial, irreverent professor who specialised in the study of Medusae (jellyfish) and sea slugs. Forbes pointed out that Huxley would have a chance to capture marine creatures from the top layers of the open sea by using a net towed behind the ship. These catches would include some of the most bizarre and beautiful creatures on the planet.

They could be as blue as the ocean, as pink as the sunset or as clear as water. Their bodies could be gelatinous, scaly, shell-encased, spiky or segmented. They moved by means of floating bells, bladders, sails, wings, legs, fins, or by water-jet propulsion. They could sting, ingest, choke, bite or poison their prey. They might reproduce by self-budding, dividing, egg-laying or sexual intercourse. Some were solitary, others gathered into elaborate colonies. Many were as translucent as the sea-water in which they swam, which meant Huxley could examine their internal structure without having to perform complex dissections.[27]

Moreover the morphological relationships of these sea creatures were still largely unknown. No credible common plan united this zoological farrago. The great French comparative anatomist Georges Cuvier had in 1812 dumped them all in a category he called Radiata (radial structured), alongside the three other major *embranchements* of vertebrates (backboned), molluscs (soft-bodied) and articulates (segmented). The name Radiata derived from Cuvier's belief that their vital organs were commonly 'disposed around a centre like the radii of a circle'.[28]

Since Cuvier's time, a number of biologists had criticised the category, but none had come up with anything better. The Radiata remained what Huxley called a 'sort of zoological lumber-room'.[29] The classifications needed reforming, and who better than the enterprising young Huxley to undertake the task.

Though his plan to study so many marine creatures was wildly overambitious, his work ethic could not be faulted. Huxley produced his first paper before the *Rattlesnake* left Portsmouth. Edward Forbes had given him an *Amphioxus lanceolatus* to examine – a strange vertebrate creature without a heart, which scuttled in submarine sands – and Huxley discovered that its blood composition resembled that of invertebrates. This was worth a note, subsequently read out by Forbes at the next meeting of the British Association for the Advancement of Science.[30]

Once the *Rattlesnake* reached open sea, Huxley graduated to more challenging work. Each day when the ship heaved to at noon for soundings, he would toss the tow net overboard. On most evenings he would retrieve a haul of pulsating jellyfish, sea nettles, sea squirts, arrow-worms, sea slugs and sea cucumbers. These he examined and dissected deep into the night, knowing they would be rotten by the following day.

On 17 January 1847, he triumphantly caught his first Portuguese man-of-war, or *Physalia*. With its translucent, gas-filled bladder and delicate indigo trail of stinging cells, it looked both whimsical and sinister. Within four days he'd found that previous scientific descriptions were 'horridly superficial'.[31]

In exotic Rio, he spent most of his time dredging local bays, then examining the blood and generative organs of a local species of *Amphioxus*. By 24 February, he'd observed enough of the daily haul of jellyfish, squirts and hydra to identify remarkable links between them. His engineer's brain and artist's eye could detect underlying affinities within disparate structures.

At Simons Bay, off Cape Town, he finished a paper on the structure of *Physalia*. Captain Stanley persuaded him to send this to his father, the Bishop of Norwich, who happened also to be President of the Linnean Society. Though no lover of bishops, Huxley complied, but he presumed grumpily that his paper would be printed only to foster the career of Captain Stanley rather than for its intrinsic merit.[32]

These quick-fire papers did not come without struggle. Huxley's early belief that the ship would be science-friendly soured. The work conditions were atrocious. No scientific equipment whatever had been supplied. He improvised his tow-net from ship's bunting, after being forced to gather specimens with a rice sieve or his bare hands. The ship itself, an elderly 'donkey'- or 'jackass'-class frigate built six years before Huxley's birth, was a hundred and fourteen feet long and crammed with a hundred and eighty people. It had been so poorly refitted that the lower deck leaked freely, and inches of water sloshed around Huxley's cabin floor for the whole of the voyage. By comparison, conditions on Darwin's *Beagle* and Hooker's *Erebus* were luxurious.

All Huxley's intricate observations, dissections and drawings had to be done in a corner of the chartroom, with his microscope lashed to the table to offset the *Rattlesnake*'s continual lumbering roll. The same room was home to twenty-two midshipmen, most aged around fourteen or fifteen. A zoo would have been quieter. And thanks to the Admiralty's meanness, the only reference books had been bought out of Huxley's own pocket. One of them, a century-old natural history by the Comte de Buffon, caused Huxley's guffawing adolescent shipmates to nickname the objects of his investigations 'buffons'. 'What a precious pack have I to deal with in these precious messmates of mine', he wrote acidly in his diary only a few days into the voyage. 'I shall make it my business to have very little to do with them.'[33]

In practice, though, he grew fond of these rowdy boys, and they of him. Despite his occasional grumpiness, they enjoyed Huxley's egalitarianism and readiness to join in their jokes.[34]

A deeper problem, he was to claim in retrospect, was the incommensurate cultures of work on board a survey vessel: 'The practical shiftiness required by the sailor in his constant struggle with the elements is . . . far apart from the speculative acuteness and abstraction necessary to the man of science'.[35] In other words, sailors had little time for philosophers, as Darwin had also discovered. The officers' unthinking indoctrination with naval routine and discipline generated constant petty clashes:

> if you want a boat for dredging, 10 chances to 1 they are always actually or potentially otherwise disposed of: if you leave your towing net trailing astern, in search of new creatures, in some promising patch of discoloured water, it is, in all probability, found to have a wonderful effect in stopping the ship's way, and it is hauled in as soon as your back is turned; or a careful dissection waiting to be drawn may find its way overboard as a "mess".[36]

Even the presence of a captain sympathetic to science, as Stanley claimed to be, was no guarantee of a general spirit of cooperation. While it's true that Huxley did not have the privileged relationship with his captain and officers that Darwin and Hooker did, on the basis of their social standing, Huxley's prickliness was also an obstacle to friendships with his seniors, as he proudly admitted:

> I often fancy that if I took the trouble to court [Captain Stanley] a little we should be great friends – as it is I always get out of his way and shall do so to the end of the story. That same stiff-neckedness (for which I heartily thank God) stands in the way with others, my "superior" officers in this ship, who if I consulted their tastes a little more and my own a little less would I'm sure think me what is ordinarily called "a capital fellow", ie. a great fool.[37]

Huxley's introversion and sense of superiority go a long way towards explaining his feeling of alienation on the ship, but the mode of science he practised was also isolating. He actually liked several of his colleagues, especially the botanically minded surgeon, John Thomson, and the official naturalist, John MacGillivray, who was there because Stanley, as a scientist himself, valued the presence of naturalists more than most survey captains. 'Jock', as friends called the jovial, red-bearded Scot, was exactly the kind of knowledgeable, hard-working and hard-drinking man that Huxley respected. Early in the voyage they several times gathered marine specimens together, then rounded off the day with a succession of 'sherry cobblers'.

But where Thomson and MacGillivray were collectors – a form of naturalism that was convivial and financially attractive to many officers and seamen, who hoped to sell their specimens in England – Huxley was an analyst, philosopher and theorist. His work was, by nature, solitary – 'I neither interfere with nor need the assistance of anyone else'.[38] He was a lone hunter for new laws, known to have theories on everything. When MacGillivray happened once to be stung by a corraline, Huxley even 'had a theory on that subject' and 'afterwards showed me uriticating organs similar to those so well developed in the Physalia'.[39]

This same intellectual fecundity enabled Huxley to lay the groundwork of a scientific breakthrough well before the ship reached its southern destination of Sydney. 'I have a grand project floating through my head of working up a regular monograph on the Mollusca [soft-bodied creatures] anatomy, physiology and histology based on at least one species of every genus,' he recorded in Simons Bay, Cape Town.[40] He was developing a method of analysing a single problem and then drawing wide-ranging theoretical conclusions from it.[41]

A letter of 1 August 1847 to his sister Lizzie explained the importance of his findings: 'If my present anticipations turn out correct, this paper will achieve one of the great ends of Zoology and Anatomy, viz. the reduction of two or three apparently widely separated and incongruous groups into modifications of the single type, every step of the reasoning being based on anatomical facts.'[42] Working out the true natural relations and affinities between species was a prerequisite for all further scientific advance. If one was a creationist, it would provide an accurate map of God's divine plan; if not, it was the baseline on which all larger philosophical questions depended.

On the voyage out, Huxley had discovered, by means of dissection and microscopic examination, an underlying set of structural and functional characteristics, undetected by previous naturalists. These characteristics linked several seemingly unrelated forms of Medusae (jellyfish) and the order of Siphonophora (colonies of stinging Medusae and polyps with siphon-shaped mouths), including the Portuguese man-of-war and the *Vellela* (by-the-wind sailor). Later research would add a further class of Anthozoa, including corals, to this new group.

Up until this time, all these organisms had been part of Cuvier's indiscriminate mass of Radiata. Huxley now found that they were built on a plan of 'two foundation membranes, one covering the outer surface, the other lining the stomach and its ramifications, the two separated by a . . . gelatinous mass'.[43] He named his large new phylum Nematophora, a label that indicated the common presence among them of a stinging cell.

In the event, his nomenclature was to be eclipsed by a rival term coined in a German laboratory at much the same time, and which was eventually supplanted in turn by the modern name Cnidarians, which reflects the presence of both dual membranes and nematocysts, or stinging cells. None of this, however, diminished the dazzling originality of Huxley's achievement in clarifying and reordering what had defeated the great Cuvier.

Huxley had worked out the link between these variable radiate animals by deploying the contemporary zoological concept of the 'archetype', a basic blueprint or ground plan to which seemingly diverse animals were supposed to conform. For some anatomists, like Richard Owen, this archetype had a metaphysical character. Owen regarded it as the idea operating in God's mind when he created zoological species, not unlike the ancient philosopher Plato's notion of a perfect 'ideal' type. For rational Thomas Huxley, however, whatever the archetype was, it was not God-thought. Whether such an archetype even existed anywhere, he didn't know, but he found the notion an invaluable tool for analysing and comparing anatomical structures. In practice, he acted as if the Nematophora archetype was a material reality, even though he had to imagine what it might look like.[44] The best way for us to think about it might be as a type of lowest common denominator.

Some modern commentators have assumed that Huxley was thinking along evolutionary lines here – that he was hypothesising the existence of some ancestral animal from which his varieties of Nematophora had evolved over time. But this is to jump the historical gun. Like most respectable English anatomists at this time, he regarded evolution as a crackpot idea associated with Lamarck's contentious theory that animals could will and then transmit physiological changes to their offspring.

Even though evolution now seems the obvious missing 'law' to explain how Huxley's linked set of marine creatures came into being, he had nothing but contempt for the idea. As we shall see, Darwin would have to expend a great deal of energy to change Huxley's mind.

In the southern autumn of 1847, as he paced the quarter-deck pondering these morphological problems, Huxley also toyed with another new theory. He would later set it out in a short passage towards the end of his Medusae paper. Here he postulated a startling affinity between the Nematophora and the embryo of a chicken, an idea he'd gleaned from reading the German anatomist Von Baer, who noted that the embryos of vertebrates tended to become more like each other as one moved back through their early stages. Huxley suggested that this might also apply to invertebrates, since a chicken embryo appeared to share the two foundational membranes of jellyfish.[45]

Much later, scientists would develop this insight in the biogenetic concept of 'recapitulation': namely, that an animal seems to pass through ancestral forms in its early stages of embryonic growth. Here again the young surgeon's mate had unwittingly brushed with future strands of evolutionary theory, which, though he resisted at the time, would one day come together in a single luminous understanding offered to him by Charles Darwin.

The *Rattlesnake* eventually arrived at Sydney's Port Jackson on 16 July 1847. Because the ship had to survey the harbour for a proposed dry dock, the crew knew it would be some time before they embarked on their major task of exploring northern Australia and New Guinea. As they berthed at Farm Cove, two things pressed on Huxley's mind: an excited expectation 'of obtaining news from home after seven long months of absence', and a chafing impatience to finish his half-formulated paper 'On the anatomy and the affinities of the family of the Medusae'.[46] In the event, both intentions had to be deferred, the first because of a shipping delay, the second because he fell in love.

Thomas Huxley seemed the unlikeliest of candidates for romance. Quite apart from being a workaholic, his upbringing and education had left no time for social life beyond the odd binge with student associates. He'd mixed with few young women besides his two older sisters. At the *Rattlesnake*'s stopover in Hobart, most officers had clamoured to attend a local ball, but Huxley preferred to drink with MacGillivray in front of a pub fire. He thought of the mandatory Sydney social rounds of 'calling and being called upon' as 'humbug', on a par with ship-scrubbing and painting.[47] The three balls and two dinners he'd had to attend in the early weeks proved nothing but 'toil'.

Still, for someone of his shabby background, it was novel and flattering to be scrutinised by daughters of the Sydney elite on the hunt for marriage partners. He wrote to his sister Lizzie:

> In this corner of the universe, where men of war are rather scarce, even the old *Rattlesnake* is rather a lion, and her officers are esteemed accordingly. Besides, to tell you the truth, we are rather agreeable people than otherwise, and can manage to get up a very decent turn-out on board on occasion. What think you of your grave scientific brother turning out a ball-goer and doing the "light fantastic" to a great extent? It is great fact, I assure you.[48]

Eloquent and handsome, the surgeon's mate in his dashing uniform had no difficulty finding dance partners. Sydney had uncovered a new side to himself – Tom Huxley the gallant.

Actually, Huxley was more domestic at heart than he liked to admit. He missed Lizzie and her family painfully. She had rescued him from sadness and neglect, and was the only person who believed in his talents and encouraged him to compete for academic prizes. Living with her family while completing his medical degree had been the most fulfilled period of his life. He was devastated when Lizzie's baby daughter Flory was struck with scarlet fever, then again in late 1845 when a scandal shattered the family's happiness. Something criminal was involved, though no one knows what: perhaps Lizzie's husband John Salt killed a patient, performed a botched abortion, or became implicated in the trade of body snatching. Perhaps he simply embezzled money or owed massive debts. Whatever his crime, it was serious enough for him to have to flee to the continent and then to America, under the surname of Scott.

Thomas Huxley, the only one of his family to stay loyal to the Salts, helped smuggle Lizzie and her two infants to Antwerp, fearing all the time that they were being pursued by 'spies'. From there she managed to catch a ship to join her husband in Nashville, where he later worked as a surgeon for the Confederate troops in the American Civil War.[49]

Nostalgia for Lizzie's family also accounted for much of Huxley's difficulty in accepting the rough masculine camaraderie of shipboard life. After celebrating Christmas of 1847 on board, he wrote in disgust:

> Where is the social ease, the comfort, the heartfelt kind words, the friendly influences of a home circle? It is now two years since I formed part of such a one and that one alas! was but the last ray of a happy sun, followed by a dark night of misfortunes. Oh Lizzie! Dear Lizzie . . . would that I could have been ever by you – I would have tended you, cheered you with a care passing that of a brother – For of all of us, you and I were the only two I believe who really loved and therefore understood one another. Where shall I ever find a sister like you at once endowed with more than a man's firmness and courage in adversity and yet gifted with the tenderest heart, and mind and taste capable of the highest cultivation.[50]

Huxley certainly didn't expect to find these precious traits in a young colonial girl. But then he met Henrietta Heathorn, at a small private party, and managed to dance with her at several later balls. When they at last grabbed the chance to talk alone, Tom and Nettie discovered they were twin souls:

> They talk of people being born for one another. If ever such a case happens, I think it has happened for you and I – there is such an odd similarity in many of the circumstances of our

> lives – and still more in many of our [traits?] and peculiarities as we have often noticed – Four and twenty years ago darling you and I were both raised by nurses . . . both of us had our household early overturned and scattered by misfortune – both in later years found a home in the house of a brother-in-law. – Each makes a journey into a far country and . . . each loved the other – at once and without the slightest pretensions to judgement – discretion – or any other of these considerations which characterize the doings of sensible folk.[51]

Huxley and Henrietta became engaged after less than half a dozen meetings. Though their lives had taken different courses, their affinities were as close as Huxley's Nematophora. Nettie was only a few months younger than him, and her family life had indeed been similarly troubled. Born out of wedlock to a West Indian widow, Sarah Richardson, and Henry Heathorn, a Maidstone brewer with Micawber-like tendencies, Nettie learnt to fend for herself as a young girl in England. She'd been cared for mainly by a beloved aunt, Kate, who became her equivalent of Huxley's Lizzie. For unspecified reasons, Nettie also spent a lonely two years away from her family at school in Germany.

In 1842, her father's money troubles led him to leave England for Australia. Around two years later, Nettie, her half-sister Oriana and her mother also travelled to Port Jackson, and then on to a remote bushland site at Jamberoo, ninety miles from Sydney, where Henry Heathorn had set up a primitive mill and brewery. To her delight and her mother's chagrin, they rode to Henry's weatherboard-and-shingle shack on a bullock dray, seated on sacks stuffed with maize husks.

There Nettie demonstrated the same resilience that Huxley had so admired in his sister. She fell in love with the wild 'paradise' of the Australian bush. She revelled in the challenge of pioneer living, learning to sew dresses, to churn twelve pounds of butter before breakfast, to gallop over tree stumps on the back of a stockhorse, to tend to a distant neighbour prostrated by childbirth, and to manage their rough old cook who'd been transported to New South Wales for house burglary.

When her sister married a prosperous wine merchant, William Fanning, Nettie prepared their wedding breakfast on the verandah of the Jamberoo hut. The feast consisted of bush turkey, wanga-wanga pigeons, wild duck, salted beef and freshwater crayfish. Her father's men toasted the bride and bridegroom with a bucketful of sherry. Nettie later moved to 'Holmwood', the Fannings' elegant new house in Newtown, a few miles west of Sydney, where she took on the demanding job of running the large household and looking after her sister's children. Soon after, she met Tom Huxley.[52]

Huxley quickly regarded 'Holmwood' as his home.[53] He enjoyed sitting around the fire yarning with Will and flirting with Nettie, or watching the adept way she managed the Fanning children.[54] Yet this practical colonial girl was also an intellectual who shared his proficiency in German and his love of Romantic literature. One of her girlhood ambitions had been to become 'famous as a poetess' in the style of Schiller, and she later claimed it was this that made her stand out from other, prettier girls in Huxley's eyes.[55]

They loved discussing books together – novels, poetry and philosophy especially. He was drawn to her seriousness and moral intensity: 'had not trouble and sorrow given your character a depth and earnestness beyond that of most women I would perhaps not have loved you and you would certainly not have loved me – if you had been capricious and vain and empty as most women are – with whatever passion you have inspired in me I should not have loved you'.[56]

He wrote to his mother that Nettie's looks were less important than her character:

as to complexion she is exceedingly fair, with the Saxon yellow hair and blue eyes. Then as to face, I really don't know whether she is pretty or not. I have not been able to decide the matter in my own mind . . . her personal appearance has nothing to do with the hold she has upon my mind, for I have seen hundreds of prettier women. But I have never met with so sweet a temper, so self-sacrificing and affectionate a disposition, or so pure and womanly a mind . . .[57]

Whatever his uncertainty about her beauty, Thomas Huxley had fallen head over heels in love. He recalled the 'miserable three days' of torment as he waited to see Nettie after their first serious conversation:

I was half mad, unable to apply myself to any occupation or to rest anywhere. I felt that my happiness depended on the issue of our next interview . . . What had I, a young man, poor, prospectless, I had almost said hopeless, to do with her? What right had I to disturb the even quiet tenor of her life, to give her new anxieties and undeserved cares? . . . At times I cursed myself, and then as I thought over each look and word I felt so happy in the belief that she loved me that all obstacles were forgotten. Anxiety brought on my old nervous palpitations and I became less and less fit for quiet thought. Her image was ever before my eyes waking and sleeping . . .[58]

Radical he might have been, but Tom Huxley idealised his colonial bride-to-be in the sentimental manner of the most orthodox middle-class male. She was as angelic as Lizzie. She would be his 'ark of promise in the wilderness of life'.[59] Her example would act as 'my good genius, banishing evil from my thoughts, and actions. Bless you, dearest a thousand times. You have purified and sweetened the very springs of my being which were before but waters of Marah, dark and bitter were they.'[60]

Nettie Heathorn, Huxley believed, would refine and ennoble his troubled, restless soul. Still more importantly, she would encourage and help her beloved but flawed fiancé to become that supremely selfless being, 'a man of science'.

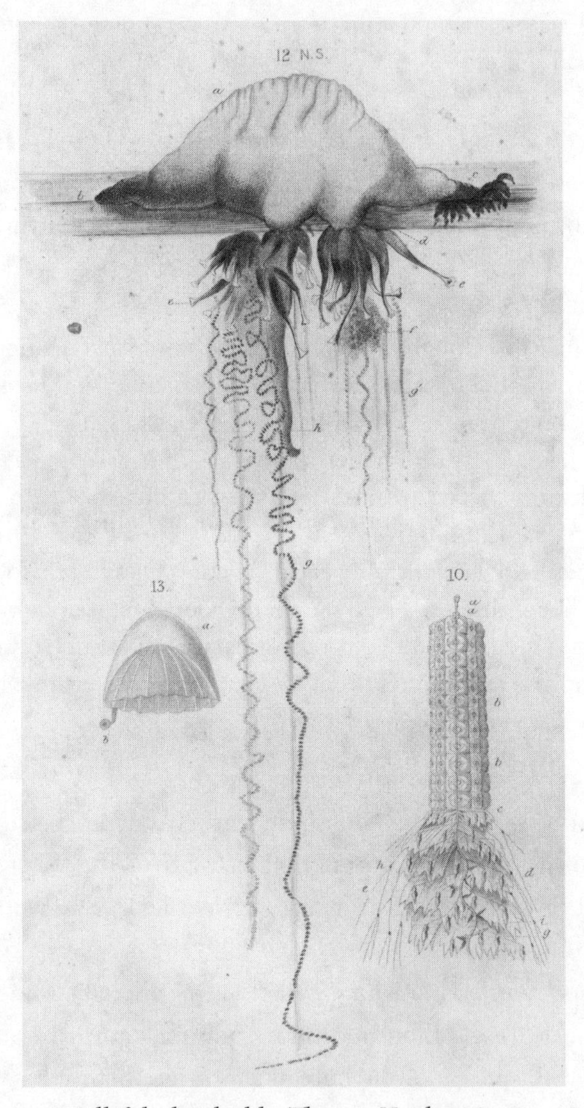


*Jellyfish sketched by Thomas Huxley, 1859*

# To Hell and Back

On Thursday 2 June 1848, Thomas Huxley rose at sunrise, dressed in his hardiest clothes, and caught a cutter from HMS *Rattlesnake* to the beach at the northern end of Rockingham Bay in Queensland, near the present-day sugar town of Mackay. He made his way to a cluster of tents, carts, sheep and horses that had been landed over the previous fortnight by a small boat through heavy swell. This was the camp of a talented young surveyor, Edmund Kennedy, who was preparing to lead a party of twelve explorers overland for some six hundred miles to Cape York, at the northern tip of Australia.

Their chief purpose was to test the viability of establishing a new military base and Asian trading port. Finding a land route to Cape York from the south would be a prerequisite for such an undertaking. Since Captain Stanley was embarking on his mission of surveying the inner route of the Great Barrier Reef at the same time, he'd volunteered the *Rattlesnake* and its smaller support vessel, the *Bramble*, to help land the expedition. The ships would also provide a rendezvous and supply point for the explorers on their final leg to Cape York.

While the *Rattlesnake* was taking on water at nearby Dunk Island, before beginning the sea-based survey, Stanley had given Huxley permission to take a short reconnaissance expedition with Kennedy, to find the best route for the land expedition to strike northward.

Huxley was excited, 'delighted at the idea of the trip and the little modicum of adventure involved in it'. After 'a capital breakfast a la bush' of tea and damper, the advance party of five men prepared to ride out. Huxley was issued with two holstered pistols, a double-barrelled carbine to sling at his side, and a cartridge belt across his chest. With his dark suntan, fierce red moustache and pointed beard, he looked, according to one friend, exactly like an Italian bandit – the type of brigand that Darwin had ridden with on the Argentine plains.[1]

Kennedy, his two trusted colleagues – John Douglas and James Luff, both ex-convicts and experienced explorers – and an Aboriginal tracker, Jacky, completed the small party. The tracker, his bush axe strapped to his saddle-bow, also led a packhorse loaded with flour to make damper, as well as sugar, tea, and a little salt pork from the ship. Huxley, thinking of Darwin's adventures in South America, decided they looked like a group of wild guerrilla raiders.

> "Are you all ready?" "All right, sir!" "Come on then." And we are fairly off into the bush. Kennedy rides ahead, ever and anon, taking a bearing with his pocket compass. I follow . . . and we two beat down the long grass into a road for those who come after us. We go along swimmingly at first but presently we come to a high ridge. We skirt first along one side, then the other, but there seems no end to it. We try to climb it, but it is composed of vile loose blocks of stone most unsatisfactory to the climbers, whether man or horse. We find it requisite to halt and Kennedy goes to the top of the ridge with Jacky and Luff leaving two of us in charge of the horses.[2]

Over the next two days, this story continued with small variations. Whatever direction they took, the path was blocked by swollen rivers, thick wooded scrub, tea-tree marshes, boggy creeks, or dense jungle where rope-like lianas impeded the horses and prickly rattans tore at their flanks. The climbing plant *Calamus australis* was the worst obstacle. As the expedition botanist explained: 'It forms a dense thicket into which it is impossible to penetrate without first cutting it away, and a person once entangled in its long tendrils has much difficulty in extricating himself, as they lay hold of everything they touch.'[3] The reconnaissance party was soaked by sudden bursts of tropical rain.

Surprisingly, Huxley didn't mind. He huddled round the smouldering fire, ate damper and drank tea from his billycan, taking his two-hour night watch with carbine at the ready, and then sleeping coiled in his cloak to keep out the driving rain, covered by a possum-skin rug to ward off the chill.

Revelling in 'the sweet discourse' with his rough companions around the fire, Huxley imagined he could do this for months. Kennedy reinforced the idea by urging him to join them on the longer expedition. Huxley seemed strong and daring, and he was also a doctor. Reluctantly the surgeon's mate had to decline, because Captain Stanley and the conditions of his commission wouldn't allow it.

Huxley little realised that in so doing he was saving his own life. The obstacles of the past few days had been prophetic of the whole trip; lethal additions of fever, starvation and native spears would turn it from frustration to tragedy.

Their first meeting with Aborigines was an omen. They stumbled across a small group who instantly vanished into the bush. 'As we rode along,' Huxley recorded, 'we heard shouts and coo-eys on all sides of us without being able to see anyone. I am free to confess I was not sufficiently used to this to feel quite easy and began to see that my pistols were handy and to speculate upon the kind of sensation that would be produced by a spear between the shoulders.' This was exactly how Kennedy would die a few months later, with a spear in his back and the bones of all but three of the party scattered around the bush.[4]

Huxley had longed to join Kennedy because he was bored and keen for adventure. As with Darwin, adventure and science went hand in hand. Huxley believed that the 29-year-old Kennedy would not only beat a new path through the Australian wilderness, he might also find animals and plants that other Europeans had never seen. Kennedy had the resourcefulness and leadership to succeed in his tough mission. 'Wholly without pretension', he was 'a fine noble fellow' with all the necessary ingredients to be a Carlylean hero. Most importantly, Huxley told Nettie in a later letter, Kennedy was 'a man of great determination (a quality of which, as you know, I am rather enamoured)'.[5]

On top of all this, joining the land expedition would have allowed Huxley to escape from the ship. 'Two or three months in the bush would have set me up in strength for the next three years and pared away the time marvellously.'[6]

His eagerness to take to the bush was in some ways surprising, because Huxley had begun the survey stage of the voyage in unusually good spirits. On 4 May 1848, four days after they left Sydney for Cape York, he'd celebrated his twenty-third birthday, in a much more upbeat mood than on his previous birthday at Mauritius. Looking back over the past twelve months, he reflected on 'what an immense change has this twenty-third year made in me! Perhaps taken all together it will turn out to be the most important of my life. My first year of sea-life – my first year of scientific investigation – the success or failure of which must determine my prospects – my first year, last but not least, of love which has and will model my future life'.[7]

Though no critic could be tougher on Huxley than he was himself, he felt satisfied with his achievements in all three of these spheres. During the past five months, he'd sailed on two short voyages from their Sydney base: to inspect lighthouses in the Bass Strait on behalf of the colonial government, and a reconnaissance trip to the Great Barrier Reef. On both occasions, he'd found himself much happier with sea life than on the voyage out. 'Formerly,' he told Nettie, 'my thoughts were something like my strides up and down the quarter deck. They were numerous without question – and cost me a good deal of invention – but they led nowhere.'[8] Nettie's love had changed all this: 'you have given my life the thing it wanted – an object to which its energies might be directed'.[9]

The effect on his mood was magical; he was getting on famously with everyone on the ship.[10] He'd also acquired at least one intimate friend in Archie McClatchie, the assistant surgeon of the *Bramble*. Because Archie was also in love – with Nettie's best friend, Alice Radford, no less – the two surgeon's mates were able to spend hours together talking about the emotional intoxication of the past five months. And if Archie wasn't around, Huxley was more than 'content with my own sweet society and that of my books, not by any means forgetting my own dear Netta's letters'.[11]

Huxley had also used those two short voyages to train himself to cope with much longer forthcoming separations from Nettie, so 'that her dear image should not be a mournful monument of past joys but rather be . . . with me in my present wanderings as a Penates'. To ensure that lovesickness did not impede his morphological studies, he disciplined himself not to read Nettie's letters or evoke her image in his mind until one hour before he slept each night.[12]

His research had prospered as a result. He reported to Lizzie in a letter from Sydney on 21 March 1848 that 'habit, inclination and now a sense of duty keep me at work, and the nature of our cruise afford me opportunities such as none but a blind man would fail to make use of'.[13] Recent excursions to Hobart, Port Phillip and Moreton

Bay had enriched the range of jellyfish and sea nettles needed to complete his major paper on the anatomy of the Medusae family.

Indeed, just before embarking on the Cape York voyage, he'd finished his Medusae paper, which he hoped would 'make a turning point' in his career. If he was right about his new class of Nematophora, he would be seen as a marine biologist, rather than a lowly surgeon's mate. It fell on few naturalists to be able to revise the work of the mighty Cuvier.[14] 'The other day I submitted it to William Macleay (the celebrated propounder of the Quinary system) who has a beautiful place near Sydney, and, I hear, "werry much approves of what I have done",' Huxley told Lizzie. 'All this goes to the comforting side of the question and gives me hope of being able to follow out my favourite pursuits in course of time without hindrance to what is now the main object of my life. I tell Netty to look to being a "Frau Professorin" one of these odd days.'[15]

He was imagining the possibility of fulfilling two simultaneous dreams: marrying Nettie and becoming a paid academic scientist. Macleay had given Huxley an enormous intellectual and psychological boost in this regard. The elderly local naturalist, a man respected by eminent English figures, including Darwin, had thrown open his own reference library, discussed and praised the Medusae paper at length, and offered, if necessary, to send it to his own circle of disciples within the zoological section of the Linnean Society.

Despite the disparities in their ages and backgrounds, the ideas of the two men were oddly, if temporarily, compatible. Macleay's 'Quinarian or Circular' thesis held that classes of animal were linked to each other in groups of five by chains of 'affinity', which 'may be represented by two circles meeting at one point, and having an altogether analogous structure at their corresponding points'. This idea that species could be organised within interconnected circles seemed plausible to Huxley at the time. At least Macleay was attempting to discover hidden affinities between species that appeared unlikely or even counterintuitive on the surface. The two men had this in common. Somewhere, Huxley believed, there could be a 'great law hidden in the circular system . . . if I could but get at it'.[16]

When it came to considering the origins of these hypothetical circular forms, however, the two naturalists parted company. To Huxley, such a system of natural classification had to be derived from an empirical law based on natural affinities or nothing. Macleay, by contrast, saw it as an expression of the mathematical harmony and symmetry bestowed by a divine creator.[17] Neither, of course, was thinking of evolution, though Huxley's naturalistic position came closer to it than did the metaphysical Macleay. Ancestral connection would in fact explain how incongruent groups of creatures could acquire underlying resemblances, though there was nothing in nature that prescribed these be in circles of five.

Both Darwin and Huxley were to abandon Macleay's Quinarian fantasy after flirting with it for a short time. While in Sydney, however, such differences of emphasis were less important to Huxley than the fact that Macleay's 'unfailing encouragement . . . gave him the courage to continue with his labours'.[18]

If Huxley's scientific work appeared to be prospering in Sydney, the coming of love into his life seemed to have effected an equally dramatic change in his personal character. Before meeting Nettie, he'd puzzled in his diary at why he always felt such a perverse refusal to enjoy even the most keenly sought achievements.

> My temperament I know too well is incapable of allowing me this so-called happiness were it ever attainable by a mortal in this world. I am content with nothing, restless and ambitious, and yet scorning the prize within my reach – the fruit of my endeavours seems to me like the fabled fruit of the Dead Sea. Once reached, it is but dust and bitterness, and I despise myself

> for the vanity which formed half the stimulus for my exertions and perhaps leads me to overestimate their value. Oh would that I were one of those plodding wise fools who having once set their hands to the plough go on nothing doubting, who so long as they only go along care not for the motives by which they are actuated nor whom they sacrifice to their selfishness.[19]

One such 'wise fool' was his friend Jock MacGillivray, who seemed limitlessly content as long as he could shoot or collect animals, birds and plants, and who was due to marry his colonial sweetheart, Williamina Gray, in just over a month's time, despite the fact that they would have little to live on. The thought of such feckless happiness made Huxley feel bitter and jealous.

The most miraculous aspect of Nettie's love, however, was that it seemed to have softened her fiancé's usual cynicism and misanthropy, enabling him at last to enjoy simple pleasures. In a letter written to his beloved 'Nettchen' soon after the trip with Kennedy, Huxley suggested that his lifelong affliction of pessimism might at last be cured. Before meeting her, he'd viewed himself as a latter-day Faust, compelled 'to stifle any such feelings as Hope'. Now he could see 'happiness dawning on the far horizon'.[20]

Nettie had even managed to inspire in 'Hal', as she'd begun to call him, something of her passionate love of Australia. When the *Rattlesnake* was making its reconnaissance to Queensland in the late autumn of 1847, Huxley took a quick inland trip from Moreton Bay, near present-day Brisbane, to the Darling Downs in the interior. He dressed himself in the typical gear of a squatter – corduroys and a cabbage-tree hat – and saddled up a hardy stockhorse. He looked exactly like the vulgar breed of Australian settler that Darwin had so abhorred, but Tom Huxley revelled in the role.

'I shall not soon forget the exhilaration of my spirits as we rode through the bush – free from all constraint,' he exulted. He scrambled his horse up the steep climb to the top of the Dividing

Range, and there experienced a moment of soaring pleasure as he looked at the scene below.

> The road shelf lay through a thick brush, with its beautiful, dense dark green foliage and whimsical festooned creepers clinging to the huge trees and thrown from branch to branch like a fantastic drapery. Here and there . . . high upon the trunk of some bare gum-tree the graceful stagshorn fern stood out like a Corinthian capital; a deep stillness reigned over all, broken only now and then by the sweet musical chime of the bell-bird . . . a few steps led to quite another scene – a wide spreading view over the distant mountains, the dense forest clothing varied in hue only by distance and the shadows of passing clouds.[21]

En route he'd also had a significant meeting with a Dr Simpson, whom he'd known slightly in England. Simpson had been an eminent exponent of the alternative medical practice of homeopathy, but had presumably grown tired of conservative resistance to his ideas. He was now working as a land commissioner and living in Woogeroo, a small town overlooking the Brisbane River, in a house that Huxley thought the prettiest he'd seen in Australia. He told Huxley that he 'was more content in his present position than any other'.[22]

This was thought-provoking for someone in Huxley's position: he teased his mother that she should not be surprised 'if you hear that I have turned colonial'.[23]

All these positive experiences of ship life, science and love over the previous year had merged to lift Huxley's spirits on the day he turned twenty-three. As the *Rattlesnake* set its course for northern Australia, he reflected that it had been a great day. Since it was now half past twelve at night, he felt entitled to begin his ritual reading of 'dear Nettie's letters'. This would complete the pleasure of an intense and satisfying bout of work. Calm conditions had attracted jellyfish. That morning, he'd caught 'a beautiful "Stephanomia" which was floating quietly along the surface of the blue mirror'.[24] He'd had a fascinating time examining this species he'd long wanted to see. Thomas Huxley was actually content.

A month or so later, things started to go terribly wrong. Without warning, Huxley tumbled into one of the most severe depressions of his life. The immediate trigger seems to have occurred just before his ride with Kennedy, but the exhilarating bush outing temporarily staved off the onset of melancholy. While the *Rattlesnake* was anchored at Dunk Island, waiting for the weather to improve so Kennedy's expedition could disembark, Huxley lay in his 'hot, wet and stinking' cabin with the rain beating incessantly down, reading a novel called *Ranthorpe* (1847), written by the English philosopher and freethinker George Henry Lewes, now better known as George Eliot's lover.

*Ranthorpe* wasn't a particularly good book, Huxley admitted, but it was so pertinent to his situation that he felt he could have written it himself. It jolted his sense of wellbeing and complacency by 'corresponding to and awakening many old thoughts'. Predictably, they were not happy thoughts. The hero of the novel, young Percy Ranthorpe, clashes with his father and throws up his solid position as a lawyer's clerk to become a poet. When his first (subsidised) publication is praised by a few reviewers, Percy becomes intoxicated with the illusion of success. Before long, he begins to posture as a literary genius. He feels himself to be another Byron – only to be disabused when his subsequent work, a Covent Garden stage tragedy, proves a humiliating flop. All his illusions are dashed. His hubristic self-deceptions have also led him to hurt his true love, Isola Churchill, an orphan who had spent time in Germany, where Percy eventually exiles himself.

Charles Darwin, 1840

ABOVE Charles Lyell, author of *Principles of Geology*, which so influenced Darwin and his colleagues

BELOW Robert FitzRoy, Captain of HMS *Beagle*

ABOVE Augustus Earle, official artist on the *Beagle*, painted this view of midshipmen's quarters on board a ship of war in 1820, before Darwin set sail with the *Beagle*, but it is typical of conditions on that ship

BELOW The still pristine beaches of the Cocos (Keeling) Islands, where little has changed since Darwin's visit. It was here that he first devised his theory on the development of coral atolls

ABOVE A South American bandit, c. 1834, similar to those Darwin encountered when he went ashore to help quell a local uprising

BELOW The residence of the chief of Kororareka, Bay of Islands, New Zealand, c. 1827. Darwin came to rue the opportunities he missed while visiting here

Joseph Hooker, 1851

ABOVE Mt Erebus, Antarctica, an active volcano discovered by the Ross Expedition and named after one of its ships

BELOW The ice ball in Antarctica, held by the crews of the *Erebus* and the *Terror* on New Year's Eve, 1842

The Kerguelen's Land cabbage plant (*Pringlea antiscorbutica*), for which Joseph Hooker produced the first thorough botanical description

A portion of Antarctica's Ross Ice Shelf, a barrier of ice the size of France that blocked the passage of the Ross Expedition to the South Pole

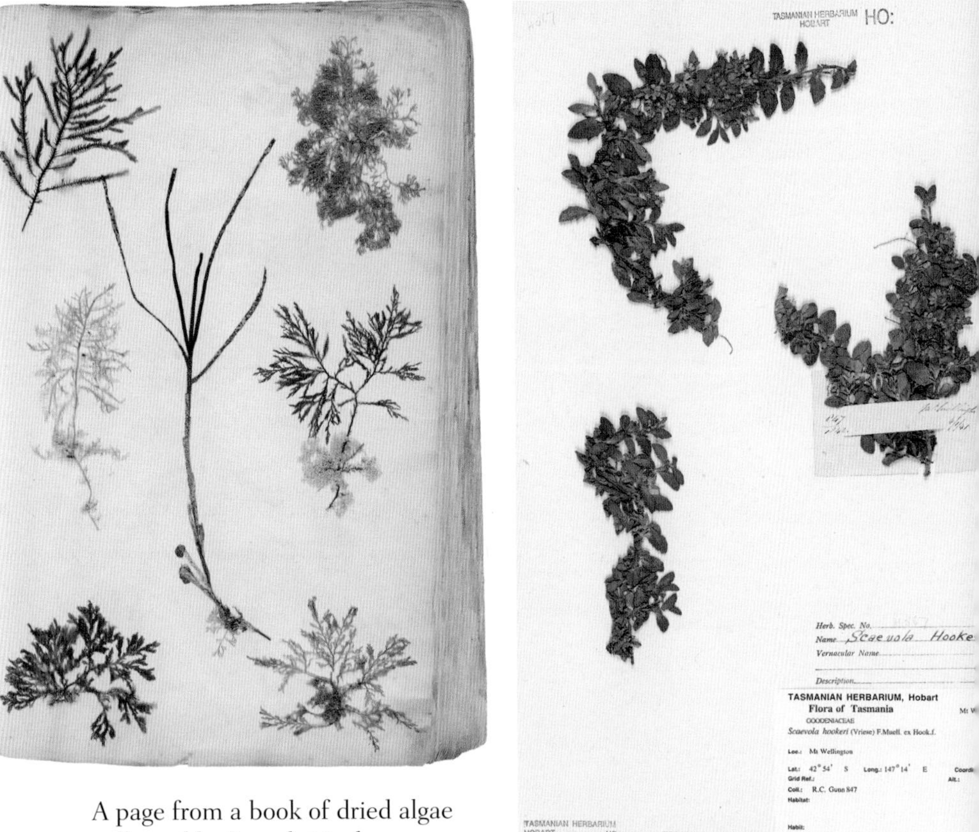


A page from a book of dried algae collected by Joseph Hooker on board the *Erebus*

A specimen of *Scaevelo hookeri* (creeping fan-flower) collected by Ronald Gunn in Van Diemen's Land

Thomas Huxley, 1857

Henrietta Heathorn, a few years after her marriage to T. H. Huxley. *From a photograph in the possession of The Hon. Mrs. John Collier.*

Henrietta Heathorn

The ball after returning from the picnic party


A ball on board the *Rattlesnake* while it was docked in Sydney Harbour

Oswald Brierly's depiction of the *Rattlesnake* entering the Louisiade Archipelago in 1849

Becalmed near the Line - "Hands to Bathe"

When the *Rattlesnake* was becalmed near the Equator, the crew took the opportunity to bathe

This sketch by Thomas Huxley, which he titled 'Ye natives finding ye Grapes are soure', was made after a skirmish on Joannet Island in the Louisiade Archipelago

Watercolour by Thomas Huxley of natives on the coast of New Guinea

to wave a spear in the air or spit on the ground and the captain acted as if they'd been threatened with a bloody battle. 'What a Sir Joshua Windbag the little man is!'[20]

Trading canoes became steadily scarcer, especially after Captain Yule – infected by Stanley's paranoia – opened fire on another harmless group of traders and killed two of them.[21] At beautiful Brumer Island, where they spent a fortnight surveying, Huxley and his colleagues were permitted to touch shore only twice, and then only for a couple of hours each time, despite the conspicuous friendliness of the inhabitants. When the *Rattlesnake* departed, the naturalists on board felt that they knew no more of the island's botany, zoology and native culture than when they'd arrived. And though the ship was anchored less than six miles from the mainland, Stanley refused to dispatch any exploratory boats. 'If this is surveying, if this is the process of English discovery,' Huxley ranted, 'God defend me from any such elaborate waste of time and opportunity.'[22]

As for the mainland itself, Stanley glimpsed enough of it from the ship to pen a lyrical description of 'the intense blue of the mountains contrasting strangely with the fleecy clouds', and to conclude from this panorama that New Guinea could prove to be the Brazil of the Antipodes.[23] Distance evidently made the heart grow fonder.

Viewing this virgin coastline should have been a 'red letter day' for Huxley also, but it only increased his frustration. 'There lies before us a grand continent – shut out from intercourse with the civilized world, more completely than China, and as rich if not richer in things rare and strange. The . . . noble rivers open wide their mouths inviting us to enter. All that is required is coolness, judgement, perseverance, to reap a rich harvest of knowledge and perhaps of more material profit. I beg pardon, that is not all that is required; a little risk is needful.'[24]

But risk was the last thing Stanley would contemplate.

'[D]iscretion is the better part of valour,' Huxley continued in a malicious parody of his captain's inner thoughts. 'And what is the advancement of knowledge . . . [compared] to my comfort, my precious life . . . keeping to the letter of your instructions earns you pay just as well.' And it was just as well that 'Cortes did not reason thus when he won Mexico for Spain, nor the noble Brooke when he conquered a province in a yacht.'[25]

Most of all, Huxley could not forgive this 'brute', coward and 'little fiend' for stopping him and his colleagues from investigating the culture of 'the human inhabitants of Papua', and so from contributing to 'the young but rapidly growing science of ethnography'.[26] This discipline had been named only a decade earlier, and it encompassed the scientific description of races or nations of men, including their customs, values and differences.

Like Darwin before him, Huxley believed that the naturalist's domain should extend to studying mankind as much as birds, rocks, plants, fish and animals. From the small contact he'd managed to make with Papuan peoples, Huxley had been impressed by 'their invariable gentleness towards each other; the kind treatment of their women; the cleanliness of their persons and of their dwellings; their progress in the useful arts, as exhibited in the pottery, cloth, cordage, pets, sails and weapons of all sorts . . . in the ingeniously built houses and canoes . . . the perseverance and design displayed in many of their carved works.'[27]

Off Brumer Island, he summed up his overall impressions: '[the Islanders] seem happy, the means of subsistence are abundant, the air warm and balmy, they are untroubled by "the malady of thought", and, so far as I can see it, civilization as we call it, would be rather a curse than a blessing to them'.[28] As a political radical and religious sceptic, his attitude to missionaries differed sharply from Darwin's. Huxley thought it 'better for the Papuans to "walk familiarly with the devils they have, than to take to themselves the seven worse, which during a long period of transition, will infallibly follow in the train of the white man, his commerce and his missionaries'.[29]

When the *Rattlesnake* left New Guinean waters to reach Cape York on 1 October, Huxley again faced the prospect of stewing on board while Stanley and Yule carried out their tedious surveys. Under the circumstances, he might easily have reburied himself in morbid fictional fantasies, particularly since he'd not long finished reading Goethe's *Sorrows of Young Werther*, a book legendary for derailing romantic young men.[30] This time, however, he decided to channel his discontent into biological study; he would take advantage of the 'friendly Buffons' swarming in the warm waters of the Coral Sea.[31]

Huxley was maturing: he could feel himself becoming more focused on achieving a scientific career, however challenging. On landing at Evans Bay in Cape York, he had a further incentive to take up his microscope. Among the letters on the provision ship was one from Professor Edward Forbes, giving 'a most satisfactory account from England of the fate of my scientific efforts'. Huxley's early papers had been read at the Linnean and Zoological societies, and his major Medusae paper was shortly due to be presented to the Royal Society. He told Nettie that the genial professor had even talked of me making 'a high name as a naturalist'.[32]

This letter to Nettie was one of the most loving he had written. He'd started to call her 'Menen', a pet name bequeathed by the Fanning children: '[N]ow and then the tears have come into my eyes as I read your loving words,' he confessed. Forbes's good news encouraged him to indulge in a bout of hopeful anticipation about their future as a married couple.[33]

Huxley's influx of domestic feelings made him think about children, which in turn led him to explore the problem of how sea squirts, 'those strange gelatinous animals', produced their offspring.[34] He'd been reading a book by a Franco-German Romantic, Adelbert von Chamisso, a young naturalist in Huxley's own mould. Chamisso had worked as a poet at the Prussian court, where he also wrote a famous romance about a Faust-like figure who sold his shadow to the devil. In 1816, he'd sailed as a naturalist aboard the imperial Russian brig *Rurik*, on a voyage of exploration to the South Seas.

During the voyage, Chamisso had formulated a widely disbelieved hypothesis about the reproductive behaviour of sea squirts. He argued that these creatures appeared initially in a solitary form, then produced offspring which were subsequently released as a long chain. Each member of this chain in turn produced a single embryo, which grew into the mature solitary form and began the cycle anew. He called this process 'alternation of generations'.

Huxley's work at Cape York confirmed this hypothesis, but also advanced a more startling claim that the two life forms of the sea squirt used different methods of reproduction. In the chain form, sea squirts used sexual organs to reproduce their solitary embryo, yet this solitary form was itself asexual. It generated its chain of offspring by a process of budding.[35]

Huxley was as much attracted by Chamisso's character as by his zoology. The poet-naturalist was a reformer who'd shared Huxley's frustrations at the boredom, lack of privacy and mindless authoritarianism of the wooden world. One of Chamisso's darker observations about his voyage on the *Rurik* struck a particular chord: 'There is something quite peculiar in life on board a ship . . . There is no means of separating oneself; no possibility of avoiding one another.'[36] Chamisso, like Huxley, hated the way his individuality was forcefully assimilated into a large and irrational naval organism.

We may hazard that it was no great jump, then, for Huxley to move from studying the reproductive system of sea squirts to reflecting on the more philosophical question of what constituted animal individuality. Should the solitary form of the salp be regarded as the biological individual, rather like the lone naturalist at sea, or should the whole chain of salps be seen as the individual, like the colony of sailors who made up a naval ship? He later extended this study to include several species from the order of Siphonophora, especially Portuguese men-of-war (*Physalia*) and by-the-wind sailors (*Vellela*). Despite himself, it seems that Huxley had found one good reason to bless the stifling lack of privacy permitted to a surgeon's mate on a British survey vessel in the tropics: it provided him with a perfect analogy for the life of a sea squirt.

What, he pondered, was the archetypal plan of such a protean creature? Was it a solitary asexual individual or a reproductive colony, or both? If the latter, how should a morphologist define or separate the creature's dual forms? These were questions that no comparative anatomist had yet thought to tackle. Huxley's answer – influenced, one suspects, by thinking about his own psychic growth on the voyage – was that the biologist in such cases should see the individual as part of an ongoing process, rather than a fixed and static state.

By analogy, Huxley himself was at one level exactly the same raw subaltern who'd set sail three years earlier, but he was also in other ways a new man, with an awakened sensibility, an intellectual maturity, and a self-confidence that he could never have previously imagined. The individual salp was similarly the complete product of its history – a succession of forms within a lifecycle that proceeded from one point of fertilisation to the next.[37]

It was a startlingly original idea. The broad conception of the biological individual as an active process – formulated in the Coral Sea as Huxley paced the quarterdeck late at night – would a few years later bring him a gold medal from the Royal Society. This was not of course an evolutionary idea, but it was a theory that recognised historical growth and development. As such, it constituted an important prelude to evolutionary thinking.

Two weeks into their stay at Evans Beach on Cape York, Huxley experienced another, wholly unexpected reason for reflecting on the malleability of the individual. 'The most remarkable occurrence that has yet befallen us happened yesterday,' he wrote in his diary on 16 October.

> A large party of natives came . . . from the islands [Prince of Wales Island in the Torres Strait] and shortly after their arrival . . . the Capt's coxn and several seamen wandering about fell in with . . . them . . . among whom was a white woman disfigured by dirt and the effect of the sun on her almost uncovered body; her face was nevertheless clean enough, and before the men had time to recover from their astonishment she advanced towards them and in hesitating broken language cried "I am a Christian – I am ashamed." The men immediately escorted her down to Heath's party ashore watering, who of course immediately took her under their protection, and the cutter arriving very shortly to take the party on board, she found herself once more safe among her own people. Three natives accompanied her off in the canoe whom she called her brothers and who appeared much interested in her.[38]

This account of the shock discovery of the castaway Barbara Thompson was broadly correct, though Huxley telescoped the sequence of events somewhat.[39] His further background to the story, gleaned from the ship's artist Brierly, was more accurate.

> Her name is [Barbara] Thompson and her maiden name was Crawford. She was born in Aberdeen and her father was a tinsmith who emigrated when she was about eight years old to Australia.

> From her account he appears to have been at first in very good business in Sydney but latterly became unsteady and consequently descending lower in the scale, was, when she left him, only a journeyman. When between fifteen and sixteen she left her father's house without his knowledge or consent and making her way up to Moreton Bay with a lover of hers was there married to him [Thompson] . . . After living . . . about 18 months at Brisbane, Thompson with his wife and three men, started in [a] cutter on their ill-omened journey [to salvage the contents of a wreck in the Torres Strait]. They had nearly reached the desired island when a heavy squall came on, and their little vessel was utterly wrecked upon a reef running out from the island. Two native canoes which were out turtle fishing were similarly distressed by the squall but the natives easily reached the shore. Not so with the unfortunate tenants of the cutter: the three men were drowned, and Mrs Thompson was drowning when one of the blackfellows [Tomagogu] swam out, and seizing her arm brought her safely to land. They treated her very kindly, fed her and protected her from insult. One of the old chiefs, who had lately lost a daughter, persisted according to their common belief that white people are the ghosts of black, that she was this very daughter "jump alive again" and she seems to have been regularly adopted among them, so that she talks of her brothers, nephews, etc. Years rolled on, and by degrees she approximated towards her friends, adopting their language so that she speaks it fluently and at present evidently thinks in it, having in talking to you to translate her native thoughts into plain English, sometimes a matter of considerable difficulty, and at the same time adopting their ways . . .[40]

The discovery of Barbara Thompson proved to be a watershed in Huxley's intellectual growth. What he learnt from her would underpin much of his later writing as a Darwinian ethnographer. This young woman of twenty-one or so – whom Huxley thought 'not bad looking' once he'd treated her inflamed eyes and burnt face – had lived with the Kaurareg people of Muralag (Prince of Wales Island) for five years. Learning their language and absorbing their ways, she had in effect become a native, in much the same way that FitzRoy's three Fuegians had temporarily become Britons. To the Kaurareg, she was Giom, a respected member of their tribal group.

As a result, she was able to open up to the naturalists on the *Rattlesnake* a world that had previously been opaque to them. Within two days of her transfer to the ship, Huxley reported that 'she has already given us a great deal of curious information about the habits of these people with an air of the most perfect truth and sincerity, and no little intelligence'.[41]

Admittedly, Giom's testimony included some insights into indigenous ways that Huxley found disconcerting. He learned how Kennedy had been killed by a dangerous inland tribe for the clothes on his back; how, when the Kaurareg killed and decapitated their enemies, they usually ate their victims' eyes and cheeks to instil bravery; and how they often killed unwanted, mainly female, babies by burying them in the sand.

On the other hand, he also learned of the Kaurareg's deep understanding of and reverence for the natural world around them. They believed that sharks and porpoises were enchanted and should be protected. Even turtles, a food staple, were seen as intricately linked with the cosmos: 'when the heavy squall of these parts gathers and the clouds topple over one another in huge fantastic masses they say the "marki" (ghosts) are looking out for turtle and they profess that one comes down and fetches a supply for the rest.' Finally, he received eye-opening glimpses into the Kaurareg's social complexity, and the importance of male initiation and death rituals.[42]

Yet these were only stray fragments of Kaurareg culture compared with the depth of information that Oswald Brierly uncovered. And it was really through Brierly's mediations that Huxley moved decisively closer to understanding the indigenous cultures of the Torres Strait. As a friend and social equal of the captain, Brierly was allowed to visit Barbara Thompson's otherwise segregated cabin and interview her at length.

Though untrained, his methods would have done justice to a modern ethnographer. He even read back his transcriptions to the illiterate young Scotswoman to ensure he'd grasped her meaning accurately. Patiently he quizzed her on rituals, myths, magic, religious values, love, birth, friendship, kinship, death, child-rearing, initiation, education, music, singing, dancing, warfare, ship and house building, sailing, hunting, craft work, food, medicine, agriculture, clothing, fashion, body scarification, and much more.[43]

The fact that Brierly and Huxley sketched together may also have heightened their appreciation of close detection. In his journal, Brierly urged the need 'to record what actually passes under your observation – any characteristic traits or circumstances which transpire under your eyes should be written down while the impression is fresh and as quickly after their occurrences as opportunity may allow – in doing this you will be constantly surprised to find the savage so utterly different from what your preconceived ideas would make him.'[44]

Under the combined tutelage of Giom and Brierly, Huxley's descriptions of indigenous life changed in both style and substance. Gone was the earlier jocularity and irony, to be replaced by a new seriousness and empathy – something much closer to what he was to call 'the . . . science of ethnography'.[45]

When on 3 December the *Rattlesnake* left Cape York to anchor off Mt Ernest Island in the Torres Strait, the fruits of this new approach became apparent. That afternoon, Huxley and Brierly took a cutter ashore and made contact with a solitary elder on the beach whom Huxley called Panooda. After accepting a gift of ship's biscuit, Panooda permitted them to walk along an inland path until they reached 'a most beautiful opening in the brush arched over by magnificent trees and so shaded and cool with such a "dim religious light" pervading it, that it looked quite like a chapel'. There they were confronted with a series of six-foot posts supporting an intricately plaited screen, in front of which stood stones painted with grotesque human faces. Huxley understood this to be a type of tomb, but he and Brierly had no time to inquire any further that evening.

The following day, they returned to the temple-like spot, equipped with sketching materials. As Panooda – 'a first-rate fellow', according to Huxley – sat quietly beside them, occasionally jumping up to view the details of their paintings with a grunt of approval, both Huxley and Brierly felt a mystical communion with their surroundings. 'The silence and gloom heightened the strange appearance of the fantastic savage monument,' Huxley recorded. 'Now and then a stray pigeon would perch over our heads or a pair of beautiful ground doves would begin their soft cooing, or a pair of ant-thrushes could be heard answering one another's loud calls from either side of the wood. I shall never forget the beauty of the place.'[46]

Increasingly at ease with his new friends, Panooda led them to another decorated screen which contained three small turtle heads and a flat board carrying ten weathered human skulls. This, he explained, was a ceremonial place where the men sat after having killed their enemies, listening to the beating of drums and undertaking 'small dances'.[47]

Through a chance physical resemblance, Brierly was rumoured in the district to be a ghost relative of Panooda's wife, Domani, and he decided to take advantage of his spectral status by asking to see her. Up until now, women had been hidden from the Europeans. Panooda conceded the justice of the request, but remained wary until assured that Huxley was 'a good man' and would bring no guns or other 'marki' (Europeans). The two sketchers then followed the old man deep into the bush and waited at a prescribed spot until he reappeared with Domani, her infant baby; a grown-up daughter, Dowai or Dowarr; and another, younger girl child.[48]

'Baba [Father],' Domani addressed Brierly shyly, 'I am . . . your child and these are your grandchildren.'[49]

Panooda, overflowing with the happiness of the moment, explained to his wife that the two visitors were making sketches, or '[minnar] pillagie – things painted', and he asked the two marki to demonstrate how this was done. While Brierly chatted with his 'relations', Huxley sketched the vivacious young Dowarr, who relished the attention, adjusting her nose ornament and giving him 'an enchanting smile'.[50] 'There was something so mild and gentle about [the family] as they sat [there],' Brierly later recorded.[51]

Huxley's final encounter with Torres Strait people, at Darnley Island between 11 and 13 December, built on his new cultural sensitivity. Here he exchanged names with a young man, 'Do-outou', who awarded him the privilege of meeting a local coskeer, or 'wife', called Kaeta. Though flattered by the attention of this 'good looking young woman', Huxley was chagrined to discover that he was obliged to shower her and her numerous relatives with gifts. When she eventually demanded to be taken on board the *Rattlesnake* to join her husband, Huxley wondered how he would be received if he were to return to Sydney and introduce Kaeta as 'Mrs Huxley'.[52] In the event, she remained at Darnley Island, apparently satisfied with her transient husband's generosity.

Soon after, the *Rattlesnake* left the Torres Strait, sailed around the edge of the Louisiades once more, and eventually arrived back in Sydney on 6 April 1850, where Thomas Huxley was reunited with the woman he hoped would one day become his true 'coskeer'.

All was not well with Nettie. She had tried to make 'Hal's' eleven-month absence pass quickly by throwing herself into her domestic and recreational routines. She galloped her horse, swam in the sea, picnicked, rowed on the river, played music, read novels, attended church, worked on her German, danced at balls, and wrote long intimate letters to her distant fiancé. Managing 'Holmwood' and the Fannings' two children had not been easy: it sometimes proved as perilous as Hal's encounters with Papuans. On one occasion, she returned home to find the house in chaos, the children unattended, and the cook, nurse and man-servant variously unconscious or belligerent from drink.[53]

Some months later, she was obliged to watch their new butler being removed in chains after it was revealed that he was a runaway convict wanted for a series of robberies.[54] And Tom certainly encountered no wild animal in New Guinea more dangerous than the tiger snake Nettie was one day forced to kill at a picnic with the children.

Through all this, it became obvious that Nettie was falling seriously ill. She began having dizzy spells, which worsened into bouts of outright fainting. The doctor prescribed a regime of stout and wine to build her up, but it made no difference. Her appetite fell off, she lost weight, grew steadily weaker, and began to experience severe pains in her face – pain so extreme she would sometimes cry out loud and remain sleepless for several nights in a row. Sometimes she felt too weak to leave her bed in the morning.

She wrote to tell Hal that the nickname he'd given her of 'little white mouse' had become more apt than ever, as she had shrunk in size and taken on a ghostly pallor. No one, least of all the local doctor, knew what was wrong, though there was no doubting the seriousness of the symptoms. Huxley himself thought she was in the grip of some type of nervous fit, not unlike the depression he'd experienced on his voyage in the Great Barrier Reef.[55]

Nettie certainly had plenty to be anxious about. Her life was being overturned around her. The Fannings, whom she loved dearly, had decided to return to England, and the thought of saying goodbye to her four-year-old niece Alice distressed her particularly. She adored the little girl and was adored in return. As she wrote to Hal, 'the love of a child is such a tender holy love'.[56] Perhaps she wondered too whether she'd ever have children of her own. She missed Hal inconsolably:

> Where are you sailing now, dear one . . . would that my will could direct the winds how soon you should return – I long actually long for the day when your Ulysses-like wanderings will be o'er but alas I have not even an idea of when this happy time may approach – Like some poor bird of passage who weary with the journey looks round for some friendly rock in the wide ocean whereon it may repose – so do I . . . vainly look for this dear spot of rest – where we shall be together.[57]

She knew that much worse was to come. The imminent prospect of what she believed would be a three-year separation weighed on her spirits like an incubus: 'dear Hal,' she wrote in the same letter, 'the uncertainty as to our future – and the certainty that our parting brings an absence twice as long again as any we have endured – adds to my grief'.[58]

She became haunted by the prospect of death, an idea exacerbated by fears that Hal would be shipwrecked or killed by savages, and by a worry that she was herself wasting away. 'To die in the course of nature is a fearful thing,' she wrote on the last day of 1849, 'but to be suddenly snatched away is a thousandfold more appalling.'[59] She was gripped with 'bitter fantasies' of receiving a black-edged sealed letter from New Guinea, and 'dreading to know what my heart surmises'.[60]

She'd been reading the biblical prophecies of Daniel and this deepened her despair. It reminded her that she couldn't even hope to be reunited with her loved ones in some future celestial paradise – her father was an unbeliever, her mother a lip-service

Christian, and as for Hal, 'I am often very unhappy about his sentiments.' Worse, she felt her own religious devotion evaporating: 'Something has come over me of late; I cannot pray as fervently as I did.' Her dreams that Hal would act as religious guide and instructor in her life 'have borne bitter fruit'.[61]

Not long after Hal's return from the north, death did burst into their lives, from an unexpected quarter. During the last few weeks of the voyage, Stanley's health had disintegrated and Thomson, the kindly surgeon, grew worried. The captain appeared to be undergoing a complete nervous collapse; for several days his mind wandered and he rambled incoherently. As well as the strain of twenty-two years of surveying in fierce climates, he was said by his family to have inherited melancholia and epilepsy.[62]

The arduous New Guinea and Torres Strait voyage had been compounded by news at Cape York that his brother Charles had died in Van Diemen's Land. Then, on arriving back in Sydney, he was informed that his father, the bishop, had also died some months earlier. On 12 December, Stanley experienced an acute epileptic fit that caused him to fall on the deck and smash his head sickeningly. At eight am the next day, he died in the arms of the surgeon's mate who'd earlier called him a vile little brute. He was buried with all the pomp the British navy could muster at St Thomas's churchyard in the suburb of St Leonards, on Sydney's north shore.

Having to tend Stanley in the last days of his life proved a chastening experience for Huxley. He read over the captain's journal and found it filled with anxiety for the health and safety of his crew. It made Huxley feel ashamed of his own selfish petulance. The death of his captain was another of the shaping experiences that Australia brought him. In the days that were spent preparing the *Rattlesnake* to return to England under the captaincy of Yule, Huxley pondered the immense changes that had taken place in the raw and angry young surgeon's mate who'd sailed into Sydney Harbour four years earlier.

He had started the voyage as a bitter, self-obsessed prig and was leaving it as a forceful young man with a double mission: to succeed in a scientific vocation, and to earn enough to marry the women he loved. The record of his profound social, intellectual and psychic transformation was contained between the covers of his journal, which he presented to Nettie shortly before embarking.

> Here is the end of a "History of Four Years", dearest. It tells of the wanderings of a man among all varieties of human life and character, from the ball-room among the elegancies and soft nothings of society to the hut of the savage and the grand untrodden forest. It should tell of the wider and stranger wanderings of a human soul, now proud and confident, now sunk in bitter despondency – how so raised above its coarser nature by the influence of a pure and devoted love . . . Could the history of the soul be written for the time it would be fuller of change and struggle than that of the outward man, but who shall write it? I the only possible historian, am too much implicated, too interested, to tell such a story fairly.[63]

The remaining time the two lovers had together while the *Rattlesnake* was being re-victualled was short. They snatched every minute in each other's company, talking, walking hand in hand, stealing kisses. They scribbled each other notes of assurance that their decision to defer marriage was inescapable and right. Three years, Huxley promised, would find him established as a scientist in Britain, or he would return to Australia to work in any capacity. In his presence, Nettie's health improved slightly, but there was no denying that she was a sick woman. Huxley could hardly bear to leave her in such a state.

Then suddenly it was time to go. The second day of May 1850 was hot and hazy, the sea as glassy as a mirror. The harbour looked like an oil painting by former *Beagle* artist Conrad Martens. A slight north-westerly sprang up and the *Rattlesnake* began to glide towards the Heads. Huxley went up onto the poop carrying the signalman's telescope and trained it on the harbourside house where Nettie was staying. She was holding up a glass directed back at him and waving a white handkerchief with her other hand. He watched until she dwindled to a tiny speck then was shut out by the cliffs of the South Head. He clambered down to the chartroom, poured himself a glass of champagne and silently drank to 'our success and happy reunion'.

Later that evening, as he took his usual walk on the quarterdeck, he saw 'the last of the land of Australia . . . a dark grey line along the horizon backed by as splendid a sky as ever the setting sun lighted up. We part friends, O land of gum trees. I have much to thank you for.'[64]

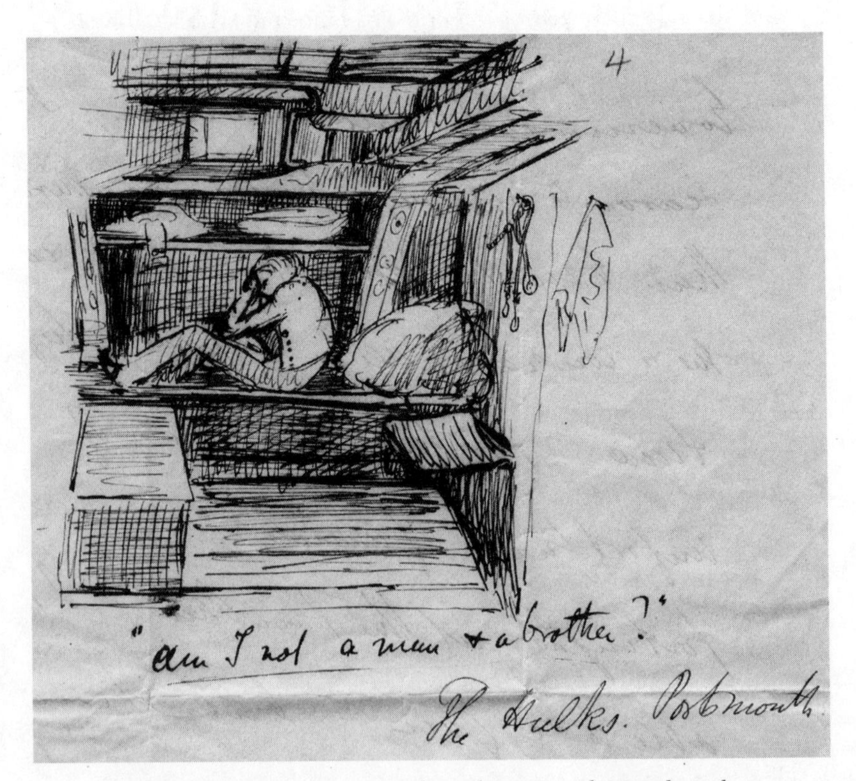


*Huxley's depiction of cramped conditions on the* Rattlesnake

# PART FOUR

# Alfred Wallace in the Amazon and South-East Asia, 1848–66

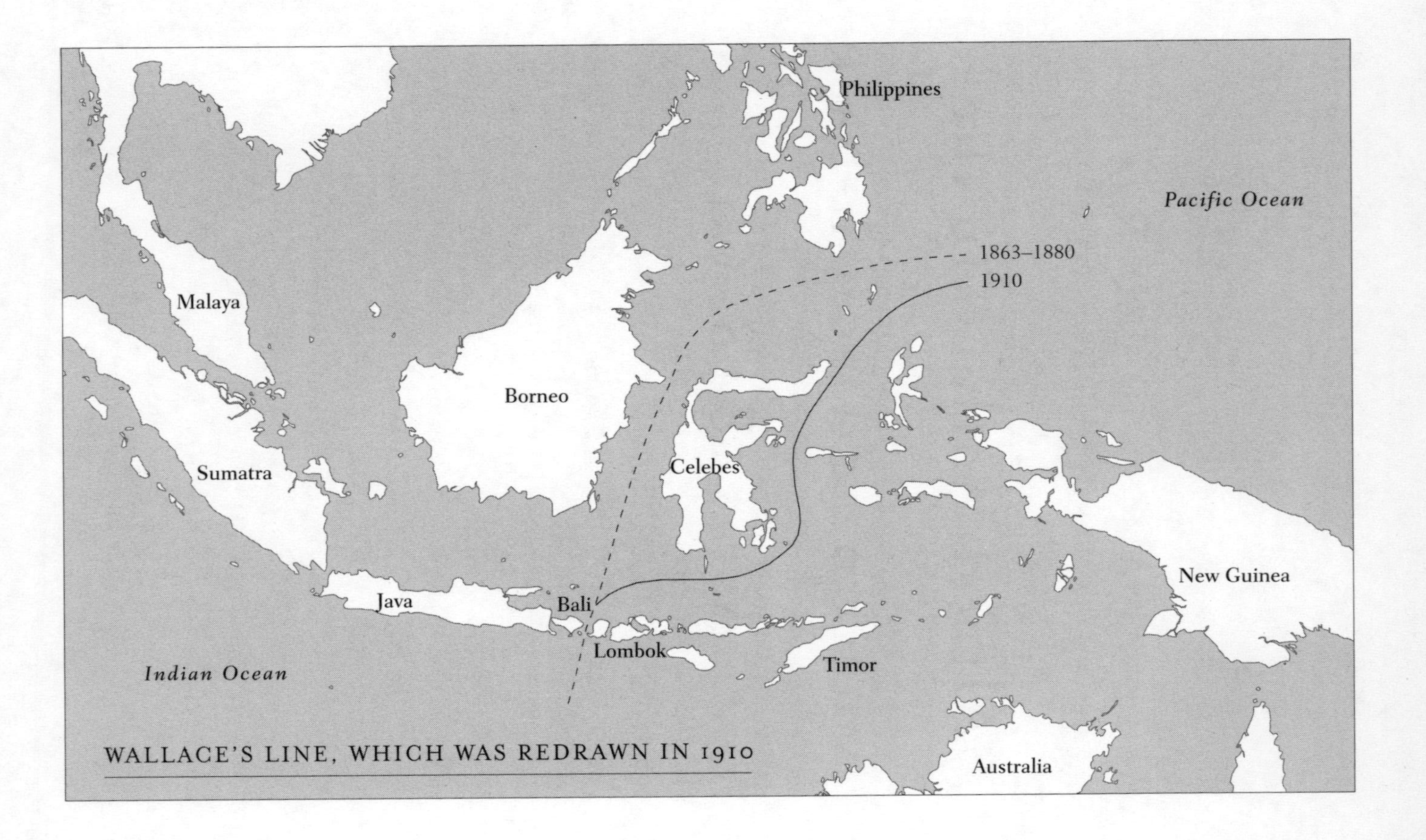

WALLACE'S LINE, WHICH WAS REDRAWN IN 1910

# A Socialist in the Amazon

At seven o'clock on a fine June evening in 1846, two brothers set off from the inn of Ystrad Fellte in the Vale of Neath, southern Wales. They walked along a sandy track and down a steep path until they reached a stream that disappeared into the mouth of a spectacular limestone cave. Known in Welsh as Porthe-y-ogof, meaning 'gateway of the cave', it was here that the river dived underground for three-quarters of a mile.

Although the water cave was a picturesque spot in daylight, with its reddish rock-face and hallway of tormented-looking stones, nightfall was an odd time to be visiting, particularly since the young men wore only thin shirts. Lacking knives, they could not cut down tree branches to make a sleeping nest, so they slumped directly on the cold stones.

The younger of the two, Alfred Russel Wallace, aged twenty-three, found it especially difficult to get comfortable. At six-foot two, his long skinny legs and bony bottom seemed to find every jagged stone. '[W]hile in health,' he later wrote, 'I have never passed a more uncomfortable night.'[1]

Despite the pain, the night in the water cave proved successful in its way, because the two Wallace brothers, John and Alfred, had gone there not to enjoy but to test themselves. As Alfred later explained in his autobiography, 'we had both of us at this time determined, if possible, to go abroad into more or less wild countries, and we wanted for once to try sleeping out-of-doors, with no shelter or bed but what nature provided'.[2]

John Wallace, who would eventually migrate to the Californian goldfields, was probably dreaming of making his fortune in the American Wild West. His brother Alfred's imagined destination was wild in a more literal sense: he wanted to work as a naturalist in the jungles of the Amazon.

Alfred Wallace's ambition appeared ludicrous. From the water caves of Wales to the jungles of the Amazon Basin was a long march. And though he came from a background similar to his slightly younger contemporary, Thomas Huxley – at that moment preparing to embark for the South Seas – Wallace was unqualified to emulate him. Wallace's father, like Huxley's, had been educated in the professions and fallen on hard times, but Thomas Wallace had fallen harder and further than George Huxley.

Alfred's father had trained as a solicitor but never practised; instead he squandered an inheritance on hopeless literary and business ventures. Successive failures forced him, his wife Mary Anne and their brood of nine children to uproot themselves several times in search of cheaper living conditions. The first of these moves, to the village of Usk in Monmouthshire on the Welsh border, coincided with the birth in 1823 of Alfred Russel Wallace, the second youngest of their surviving children. Another small inheritance – this time Mary Anne's – halted their drift into poverty just long enough for Alfred to scratch a few years of mediocre grammar schooling in Hertford.

At school, too, he bore the stigma of poverty. By the time his parents terminated Alfred's formal education at the age of thirteen, he'd been made to work as the school's only pupil-teacher in exchange for fee reductions. Naturally shy, he was so deeply cut by the humiliation that it became the source of nightmares for the rest of his life.[3]

Even before the death of Thomas Wallace in 1843, the family had been sliding down the social ladder. Mrs Wallace took a job as a household servant, and most of the Wallace boys had to opt for careers within the manual trades, at a time when Britain's industrial working classes were experiencing economic distress. The eldest son, William, managed to qualify as a surveyor, John became an apprentice carpenter, Fanny moved to America to work as a primary-school teacher, and Herbert, the youngest, became a trunk-maker's apprentice, though he longed to be a writer and a poet. Alfred himself worked for a year as an apprentice watchmaker, then became apprenticed to his surveyor brother, William.

The change in Alfred's circumstances at the age of fourteen might seem a colossal setback for a young man who'd expected to carve out a career similar to his father's, yet he never saw it that way. Working as a tradesman's apprentice turned out to be a transformative experience, one that set him on his eventual path to becoming an Amazonian naturalist. Wallace might have lost his formal education, but he was tumbled into a working-class world that prided itself on traditions of craft, skill, intellectual self-improvement, and fierce social independence. It was to mould his mind and character permanently.

Before joining William as an apprentice surveyor, Alfred, at the age of fifteen, tramped to London to live for several months with his carpenter brother John. Here he found himself plunged into a vibrant popular culture that was both a cause and a consequence of Britain's industrial revolution.

One strong strand of that culture entailed the spreading of scientific knowledge to the mechanic trades. John Wallace encouraged his younger brother to enjoy the comradeship of his small local carpenter's workshop, and he also introduced the boy to a recreational working-men's 'club' called the Hall of Science. It proved to be an institute of cooperative self-improvement run by 'advanced thinkers among workmen', and met at John Street, just off Tottenham Court Road. Most of its members were followers of the Welsh mill owner, philanthropist and socialist thinker, Robert Owen (not to be confused with the anatomist Professor Richard Owen).[4]

Alfred was electrified by the experience. His naturally idealistic personality responded to the visions of utopian change advanced by Owen and the variety of working-class radicals who lectured there. As Wallace explained in his autobiography, 'Owen contended, and proved by a grand experiment, that environment greatly modifies character, that no character is so bad that it may not be greatly improved by a really good environment acting upon it from early infancy, and that society has the power of creating such an environment.'[5]

Using innovative educational methods centred on subjects such as mathematics and science, this self-made businessman had shown that a wild tribe of Highland workers at his New Lanark cotton mills could be transformed into a disciplined and moral community of citizens. Now Owen was working to extend his vision of a 'New Moral World' to 'members unlimited'. If Alfred's contemporary, Thomas Huxley, grew up wanting to bash down the doors of privilege, young Wallace became imbued with the hope that society could be transformed altogether.[6]

As well as enjoying the pleasures of chatting, drinking coffee and playing dominoes at the Hall of Science, Wallace soaked up the 'secularist' or freethinking ideas espoused by many of the itinerant radical lecturers. Men like the Birmingham mechanic Frederick

Hollick argued that organised religion was an elite conspiracy designed to corrupt and control the people. Rational self-education was one way to break the insidious monopoly.[7] Wallace confirmed this eye-opening message by reading Tom Paine's pungent deistic polemic, *The Age of Reason*, which he found in the Hall of Science library, along with a tract by Owen's son attacking the terrorism of hellfire preachers. Both helped make him a lifelong sceptic of orthodox religion.[8]

On top of this, the Hall of Science lecturers introduced Alfred and his brothers to phrenology, a new popular 'science of the mind' that seemed to offer a rational substitute for the operations of God. Alfred also read one of the British bestsellers of the century, George Combe's *The Constitution of Man* (1828), which presented phrenological ideas in a way that appealed to a self-improving young socialist.[9]

Combe argued that the brain was the seat of the mind, an organ composed of distinct 'faculties' that housed all man's animal instincts, sentiments and intellectual abilities. These properties could be detected by tracing a person's topography of cranial bumps. Combe stressed, however, that inherited traits were not necessarily permanent: with proper education, bad tendencies could be inhibited and good ones cultivated. In this way phrenology, a do-it-yourself science for the common man, complemented Owen's social utopianism. Young Wallace quickly bought a small phrenological bust, learnt its relief map of intellectual and moral faculties, and began to practise reading people's characters.

Moving from London in 1838 to work as an apprentice surveyor strengthened the sinews of Alfred's newfound socialism in several ways. At Barton-in-the-clay, a village of small tradesmen and mechanics in Bedfordshire, the local tavern keeper introduced him to a repertoire of rollicking anti-clerical ballads. A local radical newspaper added Lord Byron and Napoleon Bonaparte to his pantheon of heroes – the polite classes would have found it difficult to decide which was the greater bogeyman.[10]

Much of Alfred's work entailed surveying for landlords wanting to enclose traditional common lands that supplemented the income of poor cottagers by providing communal rights to grazing ground and firewood. At the time, young Wallace assumed enclosure was necessary for progress, though the reality brought him bouts of unease. This intensified when he began to experience the plight of the peasant farmers of southern Wales, who were driven to join radical secret societies and engage in violent protests in defence of their traditional rights. Again, Wallace assumed that the Welsh peasant would eventually have to succumb to the march of industrialisation and scientific agriculture, yet he sympathised strongly with the peasant's 'desire to defend . . . fellow countrymen from what he considers unfair or unjust persecution'.[11]

Working as an apprentice surveyor also taught Alfred Wallace an array of intellectual skills and disciplines that would later underpin his evolutionary discoveries. While Darwin, Hooker and Huxley acquired their surveying knowledge on naval voyages, Wallace learnt the trade on land by tramping the British countryside carrying a rod, chain and sextant to measure distances and calculate angles. His brother William taught him to use thermometers, theodolites and microscopes, to navigate by the stars, and to map topographies, baselines and borders. As they made their triangulations among the mountains and valleys of southern England and Wales, William also instructed Alfred in the geological essentials for surveying.[12]

This stimulated Alfred to read Charles Lyell's *Principles of Geology*, which opened his mind to the operations of everyday geological forces – wind, tides – over deep time. It proved an essential prerequisite to understanding both evolution and biogeography.

The beauty of the countryside he was surveying inspired Alfred to take up the fashionable hobby of botany. He began collecting plants in his spare time and learning how to identify, classify and label them.[13] Before now, Alfred hadn't even known of the existence of systematic botany.[14] And though William disapproved of the time taken up by this distraction, Alfred was too excited to care.

He launched his formal botanical education by purchasing a cheap pamphlet on Linnean classification, aimed at workers and published by the Society for Diffusion of Useful Knowledge. So intriguing was the subject that he then invested the alarming sum of ten shillings and sixpence to buy Dr John Lindley's *The Elements of Botany* – the same work on natural classification methods that Joseph Hooker came to know through attending Lindley's lectures. *The Elements* explained how to classify plants by comparing the full range of their shared natural properties, using a dissecting knife and a microscope. Wallace could afford neither, but he enhanced the book's usefulness by filling every blank space with descriptions of English plants transcribed from a borrowed encyclopaedia.

Again like Hooker, he followed Lindley's advice to begin a herbarium. Every Sunday, Wallace hiked in the mountains and parks to collect specimens, which he dried between boards. 'At such times,' he recalled, 'I experienced the joy which every discovery of a new form of life gives to the lover of nature.'[15] Alfred Wallace had embarked on a new apprenticeship as a naturalist explorer.

In 1844, William Wallace's surveying business contracted sharply due to a trade recession and spiralling food prices, the latter inflated by the Corn Laws that subsidised the landed aristocracy. Forced to look for alternative work, Alfred managed to find a job as a teacher of mapping, drawing and elementary mathematics at a small school in Leicester.

In his spare time he turned for intellectual stimulation to the local Mechanics' Institute and circulating library. Self-improving educational institutions like this were springing up all over Britain during the first half of the nineteenth century, most started by businessmen who hoped to instruct the industrial working classes in practical skills, work discipline and respectable moral values. But those who attended were rarely content to have a curriculum imposed on them. Fuelled by the men's curiosity and social aspirations, the institutes often functioned as centres of independent and unorthodox ideas. In this respect, they filled an even more transformative role for scientifically minded working-men like the Wallace boys than did the universities and colleges attended by Darwin, Hooker and Huxley.

It was probably at the Leicester Mechanics' Institute that Wallace came across *A Treatise on the Geography and Classification of Animals*, written by William Swainson, an ex-army officer, naturalist and explorer. Swainson was also a talented populariser of the theories of William Sharp Macleay, the naturalist and philosopher whom Thomas Huxley had just befriended in Sydney. Wallace was more impressed by the book's division of the earth into five great biogeographic regions – Europe, Asia, America, Africa and Australia – each filled with suitably adapted animal, plant and human species. The radical freethinker in him scoffed, however, at Swainson's claim that this reflected a divine plan. 'Ridiculous', he scribbled in the book's margin.[16] Like Huxley, Wallace's first instinct was always to look for a natural law to explain zoological and botanical patterns.

Wallace moved on to read the Leicester library copy of Thomas Malthus's *Principle of Population*, drawn to the work mainly because Robert Owen disputed its argument. Wallace agreed with his guru that Malthus placed too much emphasis on the implacable law of overpopulation, but he couldn't help being impressed by the man's ideas and methods. *Principle of Population*, Wallace wrote, was the first book of 'philosophical biology' he'd ever read, and it seemed 'a masterful summary of facts and logical induction'.[17]

Malthus's thesis that overpopulation led to diminishing resources and declining wages must have cut close to the bone. Alfred had grown up seeing his own large family continually outstrip their income, and he'd also been stung by the relentless impoverishment of the southern Welsh peasantry. Malthus's ideas might be needlessly pessimistic, but they couldn't be ignored.

Along with these two vital books, the Leicester library brought Wallace a friendship that helped change his life. At a free lecture there one evening, he met Henry Walter Bates, a young man around his own age who was a product of Britain's burgeoning industrial culture of small businessmen and mechanics. His father, 'Honest Harry Bates', a Unitarian rationalist and self-taught intellectual himself, ran a small hosiery business which he expected his eldest son Henry to take over.[18] Henry had been made to leave school at twelve to learn a trade in preparation for this future role. Naturally bookish, he'd consoled himself for the tedium of sweeping warehouses by reading his way through the natural-history section of the Leicester library. His recreational passion was entomology, and his rich collection of local butterflies and beetles was a revelation to his new friend.

Bates inspired Wallace to begin scouring the overgrown local parks and riverbanks for insects as well as plants. More importantly, the two self-taught boys became intellectual soul mates, eagerly debating and discussing any new book of natural history they could lay their hands on. Both read von Humboldt's massive *Personal Narrative* of his travels in the Amazon, and then moved on to Charles Darwin's *Voyage of the Beagle*. Wallace was struck most by Darwin's unpretentious and accessible language. 'His style of writing I very much admire,' he wrote to Bates, 'so free from all labour, affectation, or egotism, and yet so full of interest and original thought.'[19] Here was a model for emulation.

The two opinionated young men didn't always agree. Bates was never a socialist like his new friend, and he showed no interest in another unorthodox body of ideas that set Wallace's mind on fire at this time. Around 1844 or 1845, he attended a lecture at the Leicester Mechanics' Institute by a visiting Owenite, Timothy Hall, who was demonstrating a new 'mental science' of phreno-mesmerism. A precursor of modern hypnotism, this grafted Combe's phrenological map of the mind onto an older body of 'animal magnetic' theory, expounded in the 1780s by Anton Mesmer, a Viennese physician. Mesmer had claimed that an invisible fluid circulated through all matter, including the human body, which would sicken if any blockages impeded the flow. Trained mesmerists could function like human magnets to stroke away these blockages, restore the fluid's circulation, and with it the mental and physical health of their patients.

Though tainted by associations with French revolutionary quackery, mesmerism had attracted followers among the northern English working class. Now the new hybrid of phrenology and mesmerism promised to revitalise both body and mind by means of therapeutic trances.[20] Because of its democratic accessibility, this do-it-yourself psychology appealed to self-taught thinkers like Wallace who were operating completely outside Britain's traditional or elite educational systems.

Never afraid to try things for himself, Wallace began using a map of the mind to practise mesmerising his pupils. By touching prescribed spots on the heads of suggestible boys, Wallace believed he could generate associated emotional and psychic states, ranging from sentimental and thoughtful reflections to aggressive, buffoonish and drunken antics. 'These experiments . . . convinced me, once for all,' he wrote in his autobiography, 'that the antecedently incredible may nevertheless be true; and, further, that the accusations of imposture by scientific men should have no weight whatever against the detailed observations . . . of other men, presumably as sane and sensible as their opponents.'[21]

Phreno-mesmerism supercharged Wallace's Owenite socialist beliefs because he regarded the new 'science' as a source of educational self-improvement and social reform. And of course he expected that self-interested clergymen and doctors would be hostile to such a liberating practice. In effect, Wallace had acquired all the typical characteristics of a self-taught working-class radical – boundless curiosity, reverence for reason, and a willingness to make independent judgements – but he also had a predilection for any theory that arose outside orthodox educational institutions or was disparaged by social elites. These traits were to make him both a fearlessly original and a perversely maverick thinker.

Wallace's enlivening self-education in Leicester stopped in late 1845, when William died suddenly from pneumonia, exacerbated by travelling in an open third-class rail carriage. For the surviving Wallace children, it was like undergoing the loss of their father again. All of them had relied on William's good sense and financial assistance, none more than Alfred.

In an effort to fill their brother's shoes, Alfred and John decided to try to revive William's surveying business in the southern Wales town of Neath, where he was buried. In the tough economic conditions of 'the hungry forties', William had been suffering badly from defaulters. When the two younger brothers failed to do any better at extracting payments from bankrupt Welsh farmers, they diversified into building. Here they were lucky. A former friend of William's wanted a Mechanics' Institute and free library built in Neath. Undeterred by their lack of experience, the two consulted a couple of library books on architecture and construction and set to work.

That Alfred Wallace should have helped design, build and eventually teach at the Neath Mechanics' Institute was apt in the deepest sense. The building's unpretentious elegance – evident to this day – symbolises the confidence and versatility of someone who was himself the proud product of a working man's education.

While Wallace was designing the building, his employer lent him a controversial new bestseller called *Vestiges of the Natural History of Creation* (1844). Reading it was probably the single most important intellectual experience of Wallace's life. Though published anonymously, *Vestiges* had been written by Robert Chambers, a talented Scottish journalist of radical and freethinking views who managed to blend wild speculations with an innovative natural history of the earth.

As the title hinted, the book rejected the literal version of the biblical creation story in favour of a materialist interpretation of the origin of the firmaments, of organic life on earth and global geological change. Though he carefully hid any signs of religious subversion, Chambers gave the divine creator responsibility only for setting in train nature's laws of progressive development.

And while he didn't use the term 'evolution', he sketched a historical theory of the biological interconnection of species and 'the progress of organic life upon the globe' that implied an evolutionary law.[22] New species, Chambers argued, came into existence not gradually, as the discredited French theorist Lamarck had thought, but in sudden natural jumps – rather like what is today called 'saltation'. Though he knew nothing of genetics, Chambers also suggested a link between man and apes at the early stages of their embryological development.

When Wallace wrote to Bates in Leicester asking his opinion of *Vestiges*, Bates echoed respectable scientific opinion by calling it a work of 'hasty generalization'. This was mild criticism compared to that of most naturalists, who were lashing the book in reviews. Darwin, one of the less scathing, was nevertheless troubled that it had worsened his own chances of publishing on the subject of the origin of species. *Vestiges* offered no credible mechanism of how evolution operated to create new species, but it had muddied the waters.

Joseph Hooker enjoyed the lively read and was amused by the book. Thomas Huxley called it an outright 'fiction'.[23] Alfred Wallace, with no credibility to lose, showed his characteristic independence and relish for argument by disagreeing with them all. He told Bates in a letter in 1847 that the book's central hypothesis was supported by the recent findings of von Humboldt, and by the theories of a freethinking London doctor and zoologist, William Lawrence, whose lectures on the nervous system had been released in pirated radical editions. *Vestiges*, Wallace proclaimed, was actually 'an ingenious hypothesis strongly supported by some surprising facts and analogies, but which remains to be proved by more facts and the additional light which more research might throw on the problem'.[24]

It's difficult to exaggerate the importance of *Vestiges* on the growth of Wallace's mind, which was by now well primed for its subversive breakthrough. With the wonderful hubris of a radical autodidact, he decided to travel to the Amazon to earn his living as a specimen collector, and at the same time search for the facts needed to prove the 'concept of evolution through natural law'.[25] His letter praising *Vestiges* urged Bates to join him in this daring enterprise. Together they would leave behind the troubled social and political battleground of 1840s Britain to work on the precarious bottom rung of the naturalist hierarchy in a country that remained largely unexplored. There they would look for the evidence and explanation for a law of evolution that almost every respectable British naturalist scorned.

Henry Bates was easily persuaded to join his reckless friend, but they had somehow to give the visionary plan a practical basis. Bates's father offered them a small loan to supplement their savings. They then dashed to London to try to raise more finance, as well as to study the large, if disorderly, array of Brazilian specimens at the British Museum. They were reassured to hear from the curator of butterflies,

Edward Doubleday, that the unexplored northern regions of Brazil could furnish enough rare specimens to make their trip pay.

Next they called at Kew Gardens, to find amiable Sir William Hooker equally supportive. Their youthful eagerness must have reminded him of his son Joseph, who'd only just returned from exploring the southern oceans. Sir William wrote them references and introductions, and promised to buy any unusual botanical specimens they found. Finally, the curator of the India Museum, Thomas Horsfield, gave them invaluable advice about how to preserve and transport delicate insect and bird specimens.

An element of luck accompanied these frantic preparations. To their surprise they bumped into one of their original sources of inspiration for going to Brazil, William Edwards, who happened to be visiting London from America. Edwards was an amateur naturalist and journalist not much older than them, whose travel book *Voyage Up the River Amazon* had made Brazil sound like a naturalist's paradise. He was, they were to find, prone to exaggeration and over-luscious description.[26] Still, he offered some valuable tips, including the suggestion they base themselves at Para, a small trading port connected to the mouth of the Amazon, which would be convenient for transporting specimens.

Their meeting with Samuel Stevens, an honest and efficient collector's agent, was most productive of all. He promised to secure buyers for their duplicate specimens, and to forward them enough money to survive in the Amazon.

Finally, after a short stay with the Bates family in Leicester, during which Alfred attempted to improve his rough skills in shooting and taxidermy, the two would-be naturalists embarked at Liverpool on 15 April 1848, the only passengers on a small trading barque, the *Mischief*, said to be 'a very fast sailor'.[27]

Some six weeks later, on 26 May 1848, Alfred Wallace and Henry Bates anchored off the coast of Brazil, six miles from the village of Salinas. Through the captain's glass they looked westward. Bates described their excitement: 'This was the frontier . . . of the great primeval forest characteristic of this region, which contains so many wonders in its recesses, and clothes the whole surface of the country for two thousand miles from this point to the foot of the Andes.'[28]

Compared with their counterparts on naval survey ships, these two penurious, inexperienced young men faced almost unimaginable challenges. They had no local support system whatever. If Samuel Stevens's offer to pay them threepence per specimen was to produce enough revenue to sustain their expedition, they needed to send thousands of specimens to Britain. And when they reached Para they were still so green they mistook the scrubby regrowth around the town for the forest itself: where they'd expected a carnival of abundance, they found eerie emptiness.

But once they adjusted to local conditions, they began to accumulate a steady stream of insects, dried plants and bird skins. By frantic hard work they managed at the end of their first two months to forward Stevens a respectable swag. Among insects alone, they sent 3635 specimens, including 553 species of butterfly, 450 of beetles, and 450 of other orders.[29]

This cruel pace became habitual. The two friends felt such pressure to maintain the flow of specimens, they couldn't help competing over collecting sites. After only six to eight months of working together, they argued and agreed to split up.[30] Wallace's own drive to find specimens was further compounded in June 1849 by the arrival of two more aspirant collectors, Wallace's youngest brother Herbert – or Edward, as he now liked to call himself – and a talented young botanist, Dr Richard Spruce.[31]

Alfred Wallace and Spruce were eventually to become lifelong friends, but Edward proved something of a handicap. A dreamy would-be poet who hated his work as an apprentice pattern-maker, Edward had agitated to join Alfred at the time of the latter's own departure, with the romantic idea of writing ballads about life on the wild Amazon. But the romance of the river palled, and Edward's mournful verses about the hardships added to the pressures on his brother.

After travelling for three weeks together up the Amazon to the village of Santarem, where they collected for several months, the brothers decided to go to Barra, at the confluence of the Amazon and the Rio Negro, while Richard Spruce pursued his own collection of botanical species back in Para. To obtain their necessary quantum of beetles, birds and butterflies, Alfred and Edward had to cadge river transport from a variety of raffish Portuguese and half-caste traders, whose schedules proved as erratic as their morals. Alfred had to teach himself to speak Portuguese and occasional Spanish, as well as the local Indian lingua franca, known as Lingoa Geral. Finding Indians willing to take them to the remoter spots in search of rare specimens proved especially difficult. Keeping them was even harder: some got drunk on their advance pay and failed to turn up, others deserted without warning in the most inhospitable places.

Too poor to purchase Western provisions, the Wallace brothers ate the same food as their Indian helpers. The staple was a type of cassava meal called *farinha*, which was generally flavoured with whatever local creature could be killed. Alfred took the cuisine in his stride, and over the next five years he tried sloth, tortoise, *jucuruju* (snake), manatee, monkey, iguana, alligator, cowfish, heron, armadillo, curassow-bird, worms and smoked red ants. Edward, whether because he was more squeamish or simply overwhelmed by the daily hardships, decided not to continue trekking into the interior with his brother, and on 31 August 1849 Alfred left him at

Barra to set off alone up the inky, unexplored waters of the Rio Negro.

In the weeks and months that followed, Alfred Wallace learned to travel with his specimens in a succession of small native canoes called *montarias* that stank of rotten fish and animal hides. Occasionally he relied on *obas*, even smaller and cruder dugouts hewn from a single log. His description of ascending the Rio Negro conveys something of the everyday hazards:

> We are falling swiftly down the river. There is a strong rapid carrying us, and we shall be dashed against those black masses just rising above the foaming waters . . . just as we seem in the greatest danger, the canoe wheels round in an eddy, and we are safe under the shelter of a rock. We are in still water, but close on each side of us it rages and bubbles and we must cross again . . . after some hours hard work, we at length reach the bank, perhaps not fifty yards from the obstacle which had obliged us to leave it.[32]

Frequently he and his guides had to haul their canoes up over the rapids, using ropes attached to trees. 'If I had birds or insects drying out, they were sure to be overturned, or blown by the wind, or wetted by the rain, and the same fate was shared by my notebooks and papers . . . it was an excellent lesson in patience to bear all with philosophical serenity.'[33] And when the wet season set in, all travel ceased: 'immense whirlpools . . . engulf large canoes. The waters roll like ocean waves, and leap up at intervals, forty or fifty feet in the air, as if great subaqueous explosions were taking place.'[34]

Wallace chose to navigate the treacherous waters of the Rio Negro mainly so as to reach the rocky haunts of the *gallo*, or 'cock of the rock'. This orange and maroon bird, 'shining out like a mass of brilliant flame', was one of the most beautiful in Brazil, especially when the males gathered in groups to perform spectacular courtship dances.[35] Such beauty and rarity guaranteed a good price back in London. For this reason, too, Wallace decided to risk journeying to the *Jurupari*, or Devil's Cataract, on the Orinoco River, to look for the legendary white umbrella bird. Among the flooded islands of the Rio Negro, he'd already captured some glossy blue specimens of this strange, crow-sized bird that could erect an umbrella-like dome of feathers over its head.[36] But the white version, if it existed at all, proved elusive. Having wasted precious time, Wallace reached the tributaries of the Orinoco in January 1851. No naturalist had ever been to this region.

But the wet season had set in and he found himself confronted with months of driving rain. In spite of the daily torrents, Wallace managed to collect a respectable tally of insects that were both new to science and different from those found on the lower Amazon. The last fact puzzled him: the environmental conditions were so similar it was difficult to see why the species should be so different.

The circumstances in which he operated during those lonely months on the Rio Negro and Orinoco would have crushed a lesser man. Alfred Wallace worked until late at night to preserve his daily catches, because the hot damp conditions turned dead birds and insects instantly mouldy – when they had not already been destroyed by maggots or swarms of predatory ants. His own body suffered in unison. He was bitten, sucked and stung by sandflies, wasps, mosquitoes, leeches and *piums* (buffalo gnats). At times, his hands turned purple and swelled so he could barely write. Tick-like chigoes burrowed under his toenails to lay eggs that had to be picked out with a needle each night. Vampire bats attacked every inch of exposed skin as he slept.

There were also more serious dangers. On one occasion he had to tramp at night in bare feet through a glutinous swamp infested with poisonous snakes. When, as he often did, he slipped off the many precarious single-log bridges into the river shallows, he risked having a tiny fish, the *Vandellia cirrhosa*, swim into his urinary tract and cause havoc with its needle teeth. Lurking alligators were ubiquitous and difficult to see in the olive-coloured waters of the Amazon or the opaque Rio Negro. One evening he encountered a black jaguar, when his rifle was loaded with useless birdshot; after eyeing him menacingly the creature decided to go its own way.[37]

The most lethal risk came from disease. Perhaps due to the rains, various types of fever flared up at every new destination. While alone on the upper Rio Negro, he was afflicted with successive bouts of dysentery and malaria. At the same time, Edward Wallace, back at base in Para, caught yellow fever as he waited to catch a ship home to England.[38] After suffering for a fortnight, Edward died on 8 June 1851. Henry Bates, also in Para at the time, tried to nurse the sad boy but nearly succumbed to the disease himself.

Meantime Alfred, who'd decided to voyage up the Uapes River in Venezuela, knew via letter only that his brother was seriously ill; it was months before he learnt of his death. The one consolation was that the tragedy revived Wallace's friendship with the good-hearted Bates.

Despite dosing himself heavily with quinine, Alfred was himself battling to survive repeated bouts of malaria. His delirium and fever would abate for a week or two, then re-emerge in a new cycle of depression, shivering and fatigue.[39] By the time he decided to head back to Para in June 1852, he felt he was bound for the same mass grave that held Edward and other yellow-fever victims.

These serious hardships aside, Wallace's four-year immersion in the gritty daily patterns of river travel had its compensations. It gave him an insider's perspective on both the habitats and the lives of Amazon peoples. He had far better opportunities to understand indigenous and foreign peoples than did his counterparts on British naval survey voyages. What Darwin and Huxley had merely sampled, Wallace experienced as part of his everyday life. His interactions with locals of every stamp – whether Portuguese traders, half-castes, black slaves or 'half-wild' Indians – made him resistant to most European prejudices about empire and race.

He observed, for example, how even the best Brazilian slave owners still infantilised their charges by depriving them of the motivation to fulfil their human potential. Using language that hinted at his future theory of natural selection, Wallace contended that 'It is the struggle for existence, the "battle for life," which exercises the moral faculties and calls forth the latent sparks of genius . . . and calls into action all those faculties which are the distinctive attributes of man.' Slaves were being deprived of a fundamental human need – the right to grow up.[40]

And it was as an Owenite utopian rather than a British imperialist that Wallace sometimes daydreamed of bringing out 'half-a-dozen working and industrious men and boys' to cultivate the fertile soils and natural biodiversity of the Amazon Basin.[41] The radical in him couldn't help comparing 'the dread miseries', 'dense towns', materialistic greed, and the moral and physical ills of industrial Britain with the saner world of the Amazon Indian village. People here appeared to be sociable, peaceable and healthy. It made him wistfully write:

> I'd be an Indian here, and live content
> To fish, and hunt, and paddle my canoe,
> And see my children grow, like young wild fawns,
> In health and body and peace of mind,
> Rich without wealth, and happy without gold.[42]

Most of the time, Wallace's assessment of Indian culture was more down-to-earth than this verse suggests. He spent weeks and months travelling in canoes, living in communal houses, or *maloccas*, eating local food and enjoying music and dance. With an ethnographic eye as keen as Brierly's in the Torres Strait he sketched Indian houses, clothing, ornaments, rituals and domestic relations.[43] On the Uapes River, where he met 'true forest dwellers' for the first time, he observed 'the wild and strange appearance of these handsome, naked, painted Indians, with their curious ornaments and weapons, the stamp and song and rattle which accompanies the dance, the hum of conversation in a strange language, the music of fifes and flutes and other instruments of reed, bone, and turtles' shells, the large calabashes of caxiri constantly carried about, and the great smoke-blackened gloomy houses'.[44] We can only guess how the Indians viewed this amiable, skinny man with long white legs, glittering wire spectacles, and a mania for catching insects.

In between the frenetic daily routine of collecting, Wallace never stopped thinking about the philosophical issue 'of evolution by a natural law', a problem he hoped to solve in the uncharted regions of the Amazon. What kind of law, he pondered, had determined the origins and descent of the myriad species that inhabited these riverbanks? Like Darwin in the southern oceans, Wallace found that the Amazon Basin provoked dozens of new questions but no clear answers. Where islands and archipelagos had led Darwin to speculate about variety and difference, Wallace observed that the network of rivers and tributaries of the Amazon created distinct regions that were not unlike islands themselves.

Indeed, the Amazon was so vast that it functioned as much like an ocean as a river: 'all the streams that for a length of twelve hundred miles drained from the snow-clad Andes were here congregated in the wide extent of ochre-coloured water spread out before us!'[45] More often than not, the river was so wide he could not see from one bank to the other.

When he did cross over the vast waterway, he noticed that the monkeys and insects were usually either different on each side or absent on one of the banks. It was as if the water served as a boundary to corral distinctive groups of flora and fauna. And the sheer quantity and diversity of species within these sub-worlds was staggering. In the Rio Negro alone, he described a hundred and sixty species of fish out of his estimated total of six hundred. In the Amazon and its tributaries as a whole, he speculated, the numbers would have to be 'immense'. What prevented such fish, he wondered, from swimming and spreading wherever they liked? Why did particular species or varieties stay in one tiny stream rather than inhabit its neighbour's waters?[46]

If, as creation theory would have it, God had fashioned species especially to suit their habitats, why were these environments delineated in such an arbitrary manner? Only a day's paddling from Santarem, a trading port on the confluence of the Amazon and the River Tapajoz, he came across a species of butterfly, *Didonis biblis*, that was absent from the region he'd just left, even though the two habitats seemed identical. The reverse held for the beautiful *Epicalia numilius* butterfly, which he saw only twice in his travels, once at Para and once at Javita – dissimilar towns two thousand miles apart.

In September 1848, while travelling from Para to the Tocatins River, he entered 'the land of the blue macaw'. He and his party shot at a flock of these superb parrots but missed because of the swiftness of their flight. Again Wallace was puzzled: 'Lower down the river they are scarcely ever seen, and never below Baiao, while from this place up they are very abundant. What can be the causes which so exactly limit the range of such a strongly-flying bird?' Perhaps, he thought, it had something to do with the rocky habitations and 'a corresponding change of fruits on which the birds feed'.[47]

This in turn raised another problem. Creationists decreed that birds had been designed so their beaks exactly suited their food sources, yet he continually came across 'different groups, having scarcely any resemblance to each other, which yet feed on the same food and inhabit the same localities, [and] cannot have been so differently constructed and adorned for that purpose alone'. He could only conclude 'that there must be some other principle regulating the infinitely varied forms of animal life'.[48]

Alfred Wallace had no idea what this principle could be, though he thought it was evolutionary in some way. He hoped to be able to find the key when he returned to Britain to work through the mass of 'facts' and materials he'd accumulated from nearly four years' grind in the Amazon.

In August 1851, he'd written home anticipating what he'd do on his return. Taking for granted that he'd produce a travel narrative from his daily journals, he proposed also writing three more specialist works: one on the myriad fish species of the Amazon and the Rio Negro, one on the assorted palms of the Amazon Valley, and one on the geography of the Amazon Basin. The last, in particular, should help to bring him closer to understanding the elusive principle regulating the rivers' carnival of diversity and specialisation.[49]

By the end of March 1852, exhausted from recurrent bouts of fever, he prepared to go home. Waiting for him at Para were most of his boxed collections, which in his absence local customs men had officiously refused to allow out of the country. After a few days of persuasion, he cleared the crates to accompany him on the return voyage. In early July, he supervised their loading onto a 235-ton brig, the *Helen*.

It was a massive collection of specimens, including ten thousand bird skins, a large herbarium of dried plants, and an unparalleled collection of birds' eggs. His personal menagerie of live animals had been greatly reduced by the hazards of the last river voyage, but he still possessed five monkeys, two macaws, twenty parrots and parakeets, five small birds, a Brazilian pheasant and a toucan.[50] Even more precious were his books of sketches and voluminous written notes and records.

Sick as he was, when Alfred Wallace boarded ship to leave the Amazon, he could be pardoned for believing that his tough four-year apprenticeship was over, and that he had reached the point where he could call himself a fully-fledged naturalist.

FREE LIBRARY, NEATH.
(*Designed by A. R. Wallace.* 1847.)
[*To face p.* 246, VOL. I.

# The Law of the Jungle

'I'm afraid the ship's on fire; come and see what you think of it.' Alfred Wallace had been reading in his cabin after breakfast on 6 August 1852 when Captain Turner popped his head through the door to make this announcement.

Wallace wasn't feeling well at the time. He was suffering from his usual seasickness, even though the winds had been light and the weather fine ever since the *Helen* departed South America three weeks earlier. His nausea was compounded by the after-effects of another bout of malaria that had struck him two days after leaving port – a bout so severe he feared he'd caught the strain of yellow fever that had 'sent my brother and so many of my countrymen to graves upon a foreign shore'.[1] Doses of calomel eventually reduced the fever, but he still struggled now when he clambered up to the foredeck to see 'dense vapoury smoke' billowing out of the focsle.

Perhaps Wallace's debilitated state prevented this most practical of men from warning against Captain Turner's next move. Turner ordered his crew to open the fore hatchway and pour water into the hold. It was a decisive mistake. By letting in 'an abundance of air', as Wallace recorded years later, the captain transformed the smouldering heat into a leaping fire.[2] Had he instead caulked every crack and sealed the hatches with tarpaulins, he would have cut off the oxygen supply and smothered the flames.

Wallace also learnt later that Turner ought to have packed wet sand around the twenty casks of a highly flammable natural lacquer called balsam of capivi, which were in the hold. Not realising that the lacquer was so volatile, the captain had used wheat chaff as a packing substitute when he ran out of sand. Now Wallace could hear the lacquer 'bubbling like some great cauldron'.[3] After its long stew in the tropical sun, the old ship was as dry as a tinderbox, and the rest of the cargo, including a hundred and twenty tons of India rubber, was already alight.

Dazed and overcome with 'a kind of apathy', Wallace shuffled back down to his cabin, which was already smoke-filled and 'suffocatingly hot'. He grabbed a couple of shirts, and a tin box containing a few old notebooks and some drawings of plants and fish, leaving behind a large portfolio of notes and sketches to join the fate of the crates of specimens now burning in the hold.

On deck, meantime, the crew scurried to make the longboat and the gig seaworthy, tossing in sails, cordage, rudders, tillers, cork bungs, navigation instruments, charts, blankets, and barrels of food and water. In the half-hour this took, the flames ate through the skylight and began scorching the quarterdeck. The captain ordered the two laden boats to be lowered into the swell, where they sank up to the gunwales and began to leak through their sun-shrunken seams. Wallace was handed a rope and told to lower himself into the gig, but being too weak to hold his own bodyweight, he burned all the skin off his palms as he tumbled into the boat. Though the saltwater seared his raw hands, he at once joined his fellow sailors in frantic bailing.

As long as the flames lasted, they hoped to attract a passing vessel. So, having tied the two boats to the stern of the *Helen* by a long rope, the refugees sat, swaying with the swell, to watch the death pangs of their ship. The fire chomped steadily through the sails, shrouds, spars and masts, until the unbalanced hulk began to roll idiotically. Helpless, Wallace had to witness his menagerie of parrots and monkeys huddle on the tip of the bowsprit to escape the heat, then turn, one after the other, to dash into the flames. The sailors rowed as close as they dared, but could rescue only one bedraggled parrot that dropped into the water from a charred bowsprit rope.

As night fell, the captain ordered the boats to move further away, to avoid the armada of burning timbers floating on the water. The poet in Wallace couldn't help admiring 'the magnificent spectacle, for the decks had completely burnt away, and as it [the ship] heaved and rolled with the swell of the sea, [it] presented its interior to us filled with liquid flame, – a fiery furnace tossed restlessly upon the ocean'.[4]

Then there was no point remaining. Having roped together the two crowded lifeboats, they moved off in a steady easterly breeze on a course they hoped would get them to the nearest landfall of Bermuda within a week.

But the island was seven hundred miles away and they hadn't reckoned on the capricious wind. In the days that followed, it first dropped to nothing, then suddenly forced them off course by swinging round to the south-east, then blew heavily from the west. Dark squalls heeled the overloaded boats, causing waves to break over their bows and undo all their bailing.

For most of the next ten days, though, it was the sun that proved their worst enemy. Their burning thirst caused the level of the water cask to drop at an alarming rate. Occasional showers of rain offered little relief, all the cloth on board being so impregnated with salt that the resultant catch proved undrinkable. Raw pork, ship's biscuit and carrots made up their daily diet. Fair-skinned Alfred Wallace alternately shivered in his thin wet clothes and sweltered as the tropical sun blistered the skin off his neck, face and hands. Like the rest of the men, he was close to despair when on 15 August they sighted a ship five miles away.

It was an old, slow, cranky brig, the *Jordeson*, bound for England and weighed down by a cargo of Cuban mahogany, fustic and other exotic timber, but it saved their lives because they were still two hundred miles from Bermuda. Captain Venables of the *Jordeson* encouraged them to drink their fill of water and generously offered to share his cabin with Turner and Wallace. 'It was now, when the danger appeared past, that I began to feel fully the greatness of my loss,' Wallace recalled.

> How many times, when almost overcome by ague, had I crawled into the forest and been rewarded by some unknown and beautiful species? How many places, which no European foot but my own had trodden, would have been recalled to my memory by the rare birds and insects they had furnished for my collection? . . . and now everything was gone, and I had not one specimen to illustrate the unknown lands I had trod, or to call back the recollections of the wild scenes that I had beheld – I tried to think as little as possible about what might have been, and to occupy myself with the state of things that actually existed.[5]

Wallace's stoicism was needed, because their ocean ordeal wasn't over yet. The *Jordeson* limped along for a week until the barometer plunged and they hit a gale. Five sails blew out and waves began to break over the bulwarks, 'making the old ship stagger like a drunken man'. Down below, Wallace was doused by water that smashed through the skylight: 'the ship creaked and shook, and plunged so madly, that I feared . . . we should go to the bottom after all'.[6]

Captain Venables shared this fear: he was snatching sleep on the settee with an axe tucked under his arm, ready to cut away the masts should they capsize. After one monster wave almost breached the ship, Captain Turner whispered to Wallace that they'd sink if hit by another, adding that he'd rather be back in their small boats than 'in this rotten old tub'.[7]

Rotten it was. A few days later, when the storm had passed, Venables had to transfer the crew's hammocks to the sail-room after fountains of water began pouring in through holes in the focsle walls where sailors had pulled away decayed wood with their hands.

Calmer seas brought other problems. Because the ship was so slow and overmanned, they twice ran short of their rotten pork and biscuit provisions and had to beg supplies from passing ships, which were going in the opposite direction. They then had to endure two further storms, the worst of which blew up in the Channel within sight of England and left four feet of seawater in their hold.

Eventually, on 1 October 1852, they docked at Deal, after a journey of eighty days, almost three times as long as it had taken Wallace to get to Brazil. Upon disembarking, all he could think about was a modest celebratory dinner: 'Oh beef-steaks and damson tart, a paradise for hungry sinners.'[8]

Alfred Wallace was an exceptionally tough and phlegmatic young man. Even so, he felt 'some need of philosophical resignation to bear my fate with patience and equanimity'.[9] He had no qualifications, no money, no patrons, no clothes. Nor did he have any publications or specimens to show for his four years of backbreaking work. His ankles were so swollen that he could barely walk, and his thin tropical shirt failed to keep out the October wind.

His agent Samuel Stevens came to the rescue. He bought Wallace a warm suit, asked his mother to house and feed the scrawny naturalist, and gave him the consoling news that his lost collection had been insured for two hundred pounds. Wallace knew his specimens had been worth twice as much, but the sum was enough to rent a house in Regent's Park for the remnants of the family – his mother, and sister Fanny and her husband – and begin rebuilding the ruins of his career.

Without theatricality, Wallace hobbled on his swollen, ulcerated legs to a meeting of the Zoological Society. Some senior members, aware of his misfortune, granted him a free ticket to visit the Zoological Gardens, and encouraged him to attend further lectures. At one of these, in December 1852, Wallace heard a young anatomist named Thomas Huxley speak about an *Echinocci* parasite that had invaded the liver of a zebra in the Zoological Gardens, causing the animal to collapse suddenly and break its neck.

Wallace had never heard a scientific paper delivered with such flair. 'I was particularly struck with his wonderful power of making a difficult and rather complex subject perfectly intelligible and extremely interesting.' Without notes, Huxley made rapid sketches and diagrams on the blackboard, showing the parasite's structure, mode of development and transformations as it migrated from the intestines to other parts of the zebra's body.

Although not long back from his Australian voyage, the erstwhile surgeon's mate seemed to have gathered an enviable reputation. Wallace couldn't believe that this seasoned professional was actually two years younger than himself. 'From that time,' he recalled, 'I always looked on Huxley as being immeasurably superior to myself in scientific knowledge.'[10]

Despite his diffidence, Wallace decided to publish the fragmentary materials he'd rescued from the Amazon trip. For anyone less resilient, this would have been a soul-destroying task. He embellished his small sketchbook on Amazonian palms by having Sir William Hooker's Kew Gardens artist, Walter Fitch, work up the plates, but he had to pay for the cost of publication himself.

Sadly, the book sold few copies. Even his botanist friend Richard

Spruce could find little to praise in the thin volume. Wallace's *Narrative of Travels on the Amazon,* cobbled together from letters home and a few early notebooks, was also slighted – by Darwin among others – for being light on facts. Yet its poor sales were probably more a reflection of Wallace's lowly status than of any significant demerit in the book, which still makes compelling reading.

Wallace also managed to deliver several papers to the Zoological and Entomological societies, on Amazonian butterfly and monkey species. Listeners were impressed by his detailed field descriptions. However, they missed the deeper implications of his comments about the geographical limits of species, even though he cited instances where the Rio Negro and Amazon formed natural boundary lines for birds, butterflies and monkeys, and drew attention to the questions this raised about how localised species came into being.[11] He pointed out, for example, that he'd collected large numbers of *Heliconidae* butterflies, making them 'among the youngest of the species, the latest in the long series of modifications which forms of life have undergone'.[12]

Alfred Wallace was implying a process of evolution, but nobody noticed because he hadn't spelled it out. Even someone as bold as he could not afford to expose himself to ridicule by publicly advocating such a heretical theory when he'd not yet accumulated the evidence to make it stick.

A less heretical paper, 'Insects Used as Food by Indians of the Amazon', had the luck to attract the attention of a powerful patron, Sir Roderick Murchison, a distinguished geologist, President of the Royal Geographical Society, and one of the best-connected scientists in Britain. A tough ex-soldier from the Highlands, Murchison was impressed by Wallace's courage, modesty, and willingness to dirty his hands in the field. The intimidating Murchison proved to be both 'accessible and kindly', to the extent of recommending Wallace for a modest government fare-subsidy in order to work as a naturalist abroad.[13]

Although Wallace had still not fully regained his health, his battered feet were beginning to itch to return to the tropics. He'd sworn not to trust himself to the oceans again, but he knew that voyaging somewhere remote was the only way to recover his lost income and credibility. His choice for 'the next scene of my wanderings' was between the Andes, the Philippines and East Africa.[14] Finally he decided to explore the Malay Archipelago, where he would again try to support himself by selling duplicate specimens.[15]

It was an inspired decision. His appetite had been whetted by reading a short entry in Goodrich's *Universal History* of 1851 that described the archipelago as 'a new world', with a vegetable and animal kingdom unlike that of any other country.[16] Later, in the opening passages of *Malay Archipelago* (1869), Wallace gave a compelling explanation for his choice:

> Situated upon the Equator, and bathed by the tepid water of the great tropical oceans, this region enjoys a climate more uniformly hot and moist than almost any other part of the globe, and teems with natural productions which are elsewhere unknown. The richest of fruits and the most precious of spices are here indigenous. It produces the giant flowers of the *Rafflesia*, the great, green-winged *Ornithoptera* [butterfly], the man-like Orang-Utan, and the gorgeous Birds of Paradise . . . To the ordinary Englishman this is perhaps the least known part of the globe . . .[17]

Even though the archipelago as a whole offered rainforests second only to Brazil's, Java was the sole island to have been worked over by naturalists. The archipelago also boasted more than four hundred volcanos and more than 17 500 islands, some isolated from the nearest continental landmass by thousands of kilometres.[18] In short, the Malay Archipelago promised to be an evolutionist's paradise, a perfect field for studying the effects of organic and geological change across time and space.[19]

Once he'd made the decision to head for the East, Wallace knew exactly how to prepare himself. This time he spent months rather than days studying and sketching the collections of the British Museum, the Linnean Society and Kew Gardens. Published reference materials proved less satisfactory, though he made one invaluable purchase. Prince Lucien Bonaparte's *Conspectus Generum Avium* was an up-to-date, 800-page catalogue of all the known species of birds and their locations, to which was added specialist information about Malayan species in the margins. A smaller volume by Boisduval describing two major butterfly species also proved an asset.[20]

Wallace decided reluctantly that he couldn't afford to collect plants, even though Sir William Hooker was keen for him to do so. Botany simply did not pay as well as zoology, which anyway was Wallace's real passion.[21] Experience had taught him that he also needed an assistant, to help with the preservation of his specimens and to act as a carer in times of illness. Fortunately Murchison secured enough money to pay not only Wallace's fare to Singapore but also that of a carpenter's son, Charles Allen, who was to be Wallace's aide.

Even now, travel crises continued to haunt Alfred Wallace. His berth on an Admiralty frigate came to nothing because the outbreak of the Crimean War early in 1854 led to the ship being reassigned as a troop carrier. Murchison again came to the rescue, with an alternative passage on a Peninsular and Orient ship, the *Bengal*. After taking a small steamer to Alexandria, then making a colourful overland trek to Suez along a desert road littered with camel skeletons, Wallace and his assistant connected with the *Bengal*, which eventually landed them in Singapore on 20 April 1854.

This booming Asian port was to be the counterpart of Para in the Amazon, the launching point for an eight-year adventure that Wallace was to call 'the central and controlling incident of my life'.[22]

It took two years of living in Singapore, Wallace was later to say, for him to begin to make sense of the dizzying richness of human life there. The streets teemed with Malay boatmen and police, Javanese sailors and servants, traders from the Celebes and Bali, Portuguese clerks from Malacca, Klings from western India, Armenian brokers, Arab bazaar keepers, Bengali grooms and Parsee merchants. The harbour was jammed with an assortment of Malay praus, Chinese junks, fishing sampans, European merchant ships and men-of-war. The chaotic skyline showed a tangle of Mahometan mosques, Hindoo temples, Chinese joss-houses and European warehouses.[23]

On a practical level, however, Wallace adapted quickly to this bewildering diversity. Within weeks he'd found ideal accommodation at the mission of a multilingual Jesuit near the edge of the jungle. From here, he and Charles were able to comb through woodcutters' clearings where timber-feeding beetles abounded. After a month of this, Wallace sent Samuel Stevens a consignment of seven hundred insect species, most of them unknown to Europeans. A less seasoned collector might have been troubled by a fresh bout of malaria, the sound of tigers coughing in the thickets, and the bloated leeches that had to be removed from his ankles each evening, but Wallace took all this in his stride, and when he ran out of water he didn't baulk at quenching his thirst by drinking the acrid soup captured in the pitchers of insect-eating plants.[24]

After spending July to September of 1854 in Malacca, he decided to focus his attention for the next sixteen months on the island of Borneo and its independent sultanate of Sarawak.[25] While in Singapore, Wallace had renewed his acquaintance with Sir James Brooke, the legendary 'White Rajah of Sarawak', whom he'd first met a year earlier at an opera in London. Brooke, who was passing through Singapore on his way back to Sarawak, urged Wallace to investigate the zoological riches of Borneo, and offered his hospitality. For someone so light on patronage, this was a tantalising offer.

The friendship that quickly blossomed between the Rajah and the socialist might at first seem unlikely. What did a shy, self-taught insect collector have in common with an aristocratic adventurer and imperialist whose warrior deeds and political skulduggery had already become the stuff of legend?

Brooke's celebrity extended to all corners of the British Raj. Five years earlier, Thomas Huxley had been reading on the *Rattlesnake* how 'the noble Brooke won a kingdom with a yacht'. It was a ripping yarn. Brooke was an Indian-born soldier whose mother kept a bullet cut from his back (some said his testicles) in a glass jar, and whose father had bequeathed him enough money to buy a schooner, with which he quelled a rebellion and won an independent kingdom.

By 1855, when Wallace met him, the White Rajah's stocks had declined. He was on the way to acquiring the black reputation that made him the model for Rudyard Kipling's story of imperial despotism, *The Man Who Would be King*, and, more ambiguously, Joseph Conrad's *Lord Jim*. Brooke had been made ruler of the eastern Borneo district of Sarawak by the Sultan of Brunei, for putting down rebellious Malay sultans in 1841, but he had subsequently alienated British parliamentary liberals by his opportunist extensions of territory and overzealous 'smoking out' of Borneo pirates.[26] By the time Wallace went to stay at his Sarawak palace, the Rajah had also lost his Byronic good looks. A severe bout of smallpox had scarred his face and weakened his body. Beleaguered and paranoid, he must have found Alfred Wallace's modesty and unabashed scientific enthusiasm a pleasant distraction.

For his part, the ingenuous young collector was entranced by Brooke's colourful personality and generosity. Wallace wrote to his mother from the Rajah's palace on Christmas Day, 1855, summing up his impressions:

> I have now seen a good deal of Sir James, and the more I see of him, the more I admire him. With the highest talents for government he combines the greatest goodness of heart and gentleness of manner. At the same time he has such confidence and determination, that he has put down with the greatest of ease some conspiracies of one or two Malay chiefs against him. It is a unique case in the history of the world for a European gentleman to rule over two conflicting races of semi-savages with their own consent, without any means of coercion, and depending solely on them for protection and support, and at the same time to introduce the benefits of civilisation and check all crime and semi-barbarous practices.[27]

Wallace was especially impressed with the relationship Brooke had forged with the 'primitive' headhunting Dyak people of Sarawak, who, he claimed, had been oppressed and enslaved under previous regimes. Many Dyaks, Wallace reported, viewed the Rajah as a type of god who had preserved their traditional way of life and liberated them from foreign despotism.[28] As a pair of social outsiders themselves, Wallace and Brooke were instinctively drawn to what were seen as the least 'civilized' of tribes. Wallace found the hospitable Dyaks even more 'agreeable' than the wild Huape Indians of the Amazon. He slept unarmed in their communal houses, looking up from his bed at rows of smoked heads hanging from the roof, and felt himself safer from crime than anywhere in Europe. 'The more I see of uncivilized people, the better I think of human nature on the whole,' he declared.[29]

Wallace also enjoyed the stimulating intellectual atmosphere of the Sarawak palace. The Rajah, who was largely self-educated yet a gold medallist of the Royal Geographical Society, had built up a major natural-history library and loved nothing better than debating the latest books, including Wallace's favourite, *Vestiges of the Natural History of Creation*.[30] In that relaxed and uninhibited company, Wallace's shyness vanished. Oddly enough, the palace was like being back at the John Street Hall of Science, or the Neath Mechanics' Institute. The Rajah's secretary, Spenser St John, recalled that Wallace 'pleased, instructed and delighted us by his clever and inexhaustible flow of talk – really good talk. The Rajah was pleased to have so clever a man with him and it excited his mind and brought out his brilliant ideas . . . our discussions were always either philosophical, or religious. Fast and furious would flow the argument . . .'[31]

One of the most memorable of these debates, Spenser St John recalled, arose when Wallace attempted to persuade Brooke and his companions 'that our ugly neighbours the orang-utans were our ancestors'.[32] Wallace had come to the right place to try to prove his case. Orang-utans abounded in the neighbouring jungles of Borneo, and the Rajah had published a respected study of their physical structure. Wallace told a correspondent in May 1855 that 'one of the principal reasons which induced me to come here was that it is the country of those most strange and interesting animals the orang-utans, or "Mias" of the Dyaks'.[33] For a decade – probably since reading the *Vestiges* – he'd been convinced that kindred species tended to coexist in time and space, implying that one of these species was likely to be the ancestral form of the other. Borneo, home of the orang-utan, could conceivably be the birthplace of man.

While some orang-utan skeletons had been examined by English anatomists like Richard Owen, much remained to be discovered about the structure and behaviour of these manlike primates. At the most basic level, zoologists disputed the number of distinct orang-utan species. Some argued for two, the Dyaks claimed three, and an expert in India had recently made a case for the existence of four separate forms. Next to nothing, too, was known of their behaviour in the wild. Wallace, with his strong interest in the geographical distribution of species, was particularly intrigued to find out why orang-utans were not found in the Sarawak district, even though it was environmentally similar to the rest of Borneo.[34]

So, after canoeing up the Sarawak River for several months, Wallace and his assistant travelled in June 1855 to a British-run coalmine at Simunjon, in eastern Sarawak. It proved an ideal spot. The sunny, open mine-clearings, surrounded by dense and gloomy jungle canopy, were a magnet for a plethora of species. Insects alone were so abundant that Wallace rented a two-bedroom house for a bout of sustained collecting. Within less than a square mile he was able to gather more than two thousand species of beetle. A different kind of prize was a butterfly so rare and elegant that he named it *Ornithoptera brookeana*, after his new friend the Rajah. It had long pointed wings with metallic-green, feather-shaped spots, a deep black velvet body and a crimson neckband.

Wallace was puzzled as to why he found only male specimens: could the female be as camouflaged as its mate was visible? He made another thought-provoking discovery of a small tree frog with elongated membranes on its oversize webbed feet. These it used as improvised wings to glide through the jungle like a bird. How could such an extraordinary hybrid creature have come into being?

But the greatest natural wonder was the orang-utan. During his fourteen months in Borneo, Wallace killed sixteen specimens of these rare great apes – nine males and seven females. Our modern sensibilities recoil from his matter-of-fact accounts of tearing through the jungle to fire ball after ball into their bodies. On 24 June 1856, for example, he shot 'a male of the largest size'.

> As soon as I had fired, he moved higher up the tree, and while he was doing so I fired again: and we then saw that one arm was broken. He had now reached the very highest part of an immense tree, and immediately began breaking off boughs all around, and laying them across and across to make a nest. It was

After examining the Galapagos plants, Hooker reported that the 'Flora of each Islet was ½ peculiar'. It seemed to have come from South America, yet 'I should not have known where to put it, supposing Galapagos not to exist'.[8] Darwin was 'delighted and astonished' with these careful results. 'How wonderfully they support my assertion of the difference in the animals of different islands, about which I have always been fearful,' he declared in 1846.[9]

Hooker had been additionally helpful because, like Darwin, he believed that botanical geography should be a proper inductive science. '[T]he first steps to tracing the progress of the creation of vegetation,' he'd told Darwin in 1845, 'is to know the proportion in which the groups appear in different localities, and more particularly the relation which exists between the floras of the localities, a relation which must be expressed in numbers to be at all tangible.'[10] Hooker was urging the use of 'botanical arithmetic' in order to calculate the incidence and proportions of particular species in any given region, and their relation to such factors as moisture and altitude. Only by such means could botany make claims to being a true philosophical science.

The rapid blossoming of their friendship was based on more than this shared scientific interest, important though that was. As Darwin himself put it, the two of them were 'co-circum-wanderers and fellow labourers'.[11] Having been voyagers to the southern oceans gave them an instant rapport; there was no place for formality or affectation between fellow sailors. Within a few months, they were visiting each other's houses and addressing each other with the familiarity of old friends.

Darwin's 'delightfully frank and cordial' sailor's greeting had struck Hooker the very first time he met him.[12] Hooker's reserve had been similarly moderated by the camaraderie of the *Erebus*; he'd made strong friendships with several of the junior officers, whom he later counselled on how to deal with grievances against Captain Ross and the Admiralty.[13] For the rest of his life, he never ceased to pine for the days of 'having my own cabin at sea'.[14]

Darwin, for his part, saw in the young botanist all the character-shaping virtues that shipboard life could offer: 'good humoured patience, unselfishness, the habit of acting for himself, and of making the best of everything, or contentment: in short . . . the characteristic qualities of the greater number of sailors'.[15] It was this sailor-like candour in Hooker that encouraged Darwin to entrust him in January 1844 with his closest secret. He confessed to having developed a belief in the mutability of species, and to having discovered a 'simple way by which species become adapted to various ends'.[16]

That 'simple way' was the action of a Malthusian-style struggle for survival – an idea that had clicked in Darwin's mind on rereading *Principle of Population* at a time when mass bread riots and hunger marches were also reminding him of the struggle for existence.[17] The admission of his beliefs to Hooker was, Darwin said, like 'confessing to a murder' or a terrible vice. But the young botanist had accepted the news unflinchingly. He replied that 'a gradual change of species was possible' and that he would 'be delighted to hear how you think that this change may have taken place, as no presently conceived opinions satisfy me on the subject'.[18]

Later the same year, Darwin sent Hooker a longer sketch of his theory of natural selection for comment, and soon after nominated him to edit his larger works of evolutionary materials, should he die before they were published.[19]

One source of their mutual trust was the lack of any religious differences between them. Hooker's weak current of Anglicanism had, like Darwin's, ebbed quietly away on the tides of the Pacific. Out of consideration for his parents and relatives, Hooker never flaunted his scepticism, but he was in reality something of an 'agnostic', even though his future friend Thomas Huxley would not formally coin that term until 1869. Darwin's own residual religious instincts had been extinguished in 1851 by the 'bitter and cruel loss' of his darling favourite daughter, Annie, to fever. She was as close to perfect a human being as he'd ever known, and Darwin wanted nothing to do with any divinity that could strike her down.[20]

Over the next decade, Hooker had become Darwin's confidante, advisor and best friend, the man Darwin could trust to keep his dangerous secret, to criticise his ideas honestly, and to help his scientific work unreservedly. Throughout this time, Hooker responded to the theory of natural selection like an open-minded critic. 'My great desire,' he later told Darwin, 'was to put every single objection as strongly as I could. I did not feel myself a dissenter or opponent to your views, so much as a non-consenter to them in the present state of knowledge . . .'[21] At the same time, he committed himself to his invalid friend unstintingly, putting at Darwin's disposal his own expanding network of colonial collectors and his unique knowledge of Southern Hemisphere botany and biogeography.

After undertaking another adventurous collecting trip on land between 1847 and 1849, to the central and eastern Himalayas, Hooker was also able to provide Darwin with botanical expertise on the vast Indian region. Despite his own chronic overwork writing his books and struggling to publish the fruits of his *Erebus* voyage on a slender Admiralty subsidy, Hooker patiently answered Darwin's endless questions about plant classification, modification and variation, even when this meant having to prepare lengthy tabulations and calculations.

As well, Hooker obtained and sent Darwin scores of specialised reference works and rare plant samples from Kew Gardens, and joined with him in conducting experiments on oceanic seed survival and transmission.[22] It was Hooker, too, who urged Darwin to master the structure and reproductive system of a single zoological species, and so prompted his crucial work on barnacles.[23]

The intellectual traffic between the two men was not, of course, only one way. Particularly after 1855, when Darwin finished his barnacle studies, their interests converged even more closely. Darwin encouraged Hooker to complete his treatise on the floras of Antarctica, India, New Zealand and Tasmania, stressing that without accurate identifications and descriptions it was impossible to understand how plants had distributed themselves across the globe. At the same time, he strengthened his younger colleague's self-proclaimed 'rabid radicalism' as Hooker tried to reduce the chaos generated by botanical 'splitters', who turned every tiny plant variation into a new species and thereby impeded accurate global comparisons of plant movements and adaptations.[24] Again and again, Darwin shored up Hooker's confidence in his 'philosophical' abilities.

It was this support that gave Hooker the self-belief to use his study of Tasmanian flora as a springboard for 'a general work on the Geographical distribution of the whole Australian flora'. '[T]his is ambitious,' he admitted to Darwin in 1855, 'but it is really the most extraordinary thing in the world.' He'd found, among other things, an 'astounding' disjunction between the plant species of the south-west of Australia and those of New South Wales.[25] How was this possible, he asked, on a single continent with no striking land barrier?

Another key biogeographical problem for both of them was how to explain the presence of common plant species on the widely separated landmasses of Australia, New Zealand, the Antarctic islands and South America – not to mention evidence that some Northern Hemisphere plants had somehow made their way to the Southern Hemisphere.[26] Darwin himself doubted that land bridges had ever linked these islands and continents, even New Zealand and Australia. Instead he favoured the possibility that seeds had been transmitted via birds and ocean currents, or by means of large-scale glaciation during an ancient ice age.

Hooker, by contrast, conjured up hypothetical former continents as easily, Darwin said, as he made pancakes. 'I am becoming slowly more convinced of the probability of the Southern Flora being a fragmentary one – all that remains of a great Southern continent,' Hooker had argued typically in 1851.[27] This difference of opinion led them to undertake joint experiments with seeds and saltwater tanks in 1855. Darwin's children were excited by the possibility of defeating Dr Hooker, whom they adored, and they were rewarded early the following year when the Down House contingent managed to get cress and lettuce seeds to germinate after twenty-one days of salty immersion.[28] This suggested that the plants had been dispersed by birds or ocean currents, rather than by land bridges as Hooker had originally thought.

However pleasing this small biogeographical triumph, there remained for Darwin the critical challenge of getting Hooker to accept that species originated by natural selection. It's true that by the time Hooker and his wife travelled to Down House for the weekend in April 1856, Darwin had noticed signs that his friend's resistance was crumbling. While Hooker's books on Indian and New Zealand botanical species, published in 1855, had continued to treat species as created and permanent, occasional passages suggested that Darwin's ideas were seeping through.

In his Indian study, for example, Hooker had written in a most Malthusian vein that plants 'in a state of nature are always warring with one another, contending for the monopoly of soil, – the stronger ejecting the weaker – the more vigorous overgrowing and killing the more delicate. Every modification of climate, every disturbance of soil, every interference with the existing vegetation of an area, favours some species at the expense of others.'[29]

More recently, Hooker had denied in print that there was 'a shred of evidence' to support the idea of the permanence of species.[30] And in 1854, he'd admitted half jocularly to his Harvard botanical friend

Asa Gray that he was becoming tainted with Darwin's heresy: 'Oh dear, Oh dear, my mind is not fully, faithfully, implicitly given to species as created entities *ab origine*.'[31] Even so, Darwin needed to know whether Hooker would actually come out and endorse a publication that exposed the heretical theory of natural selection to the gaze of the world.

It was no less urgent to win over Thomas Huxley, who had long called himself an 'active doubter' about evolution. On first meeting Huxley in 1853, Darwin had found the sarcastic young biologist decidedly unnerving. 'His mind is as quick as a flash of lightning and as sharp as a razor,' he later wrote.[32]

After three years of post-*Rattlesnake* struggles, Huxley had not been in the greatest of moods. Poor, lovesick and angry, he was taking every opportunity to flaunt his contempt for the aristocratic and clerical privilege that monopolised British science. With no job on the horizon and his leave from the Admiralty running out, he'd decided that 'the world is no better than an arena of gladiators and I, a stray savage, have been turned into it to fight my way with my rude club among the steel-clad fighters'.[33] Although he wanted to impress his scientific peers, he insisted it be on his own terms. 'I shall make my mark somewhere and it shall be clear and distinct . . . [and] free from the abominable blur of cant, humbug and self-seeking which surrounds everything in this present world,' he told his sister.[34]

Darwin's sailorly warmth and lack of social pretension had gone a long way towards disarming this *enfant terrible*. Though angry at the navy's refusal to fund his post-voyage publications, Huxley had come to value his four years of shipboard 'realities'. He recognised that he shared with Hooker and Darwin a 'masonic bond in . . . being well salted in early life'.[35] Having lived and worked with sailors meant that all three were unusually comfortable talking to 'horny-handed plebeians'. And it was only with these two southern-voyager scientists that lovesick Huxley could reminisce about his 'other home', twelve thousand miles across the ocean.

All three men had also served a common apprenticeship emptying tow nets each day, and dissecting their contents on a rolling ship – skills that proved invaluable when Darwin began investigating Cirripedia (barnacles), at the same time that Huxley was reclassifying marine squirts and stingers.[36]

At first Huxley had been only one among a number of Pacific experts that Darwin pressed for marine information and specimens. The others included conchologist Hugh Cuming, biologist James Dwight Dana, and the trio of amateur Sydney naturalists Thomas Mitchell, William Sharp Macleay and Syms Covington.[37] In Huxley's case, Darwin could reciprocate by giving him the *Beagle*'s sea squirts to examine. It hadn't taken much correspondence between the two for Darwin to realise that Huxley was in a league of his own, both as a biologist and a communicator. Even the hermit of Downe had heard word of the young man's crowd-pleasing scientific lectures to working men, and his punchy journalism in the *Westminster Review*, the organ of London's anticlerical avant-garde.

On meeting Huxley in person, Darwin had taken the unusually bold step of asking him to review his forthcoming book on barnacles. Sure enough, a year later Huxley published a long review praising Darwin's brilliant originality as 'an observer of nature on the small as on the large scale'.[38]

Darwin was impressed with the fearless way Huxley challenged Professor Richard Owen, the titan of British science, who enjoyed skewering any scientific theories that disputed divine creation. Though instinctive enemies and of different generations, Huxley and Owen actually had much in common. Both were brilliant, aggressive, self-made comparative anatomists who took a particular interest in parthenogenesis – the study of how organic individuals are generated in a female reproductive cell without fertilisation by a male. Both, too, were believers in biological 'archetypes'. But while Owen regarded the archetype as an idea emanating from the mind of God, Huxley treated it simply as a blueprint to show how the hidden similarities of species could be related to a common form. Owen's was a metaphysical approach, Huxley's a materialist one.

The two men also diverged fundamentally over how the vocation of science should be advanced. Owen had become adept at wooing Britain's aristocratic and clerical establishment; Huxley wanted a merit-based reformation to sweep away all unearned privilege and patronage. He longed for science to be seen as a moral calling greater than any religion.

The early, wary relationship between Owen and Huxley had deteriorated into an escalating series of disputes about cell reproduction, though most spectators, Darwin included, knew that they were really struggling over scientific dominance. Huxley thought Owen had been able to get away with treating 'the Natural World' as his private preserve because there'd been 'no one to tread on his heels'. But as Huxley asserted to his sister Lizzie, 'on my own subjects I am his master, and I am quite ready to fight half a dozen dragons'.[39] He enjoyed imagining Owen's capillaries bursting with rage each time he read a hostile review. Like the hero of one of his own articles in the *Westminster Review*, Huxley saw himself as a daring guerrilla leader waging war against a dark empire.[40]

Yet Huxley's reputation as a controversialist also posed a problem for Darwin, now that he was considering publishing. Huxley was not really much interested in the historical origin of species, being more concerned with biological structure than history, but he was hostile to evolution. He lumped it with all metaphysical ideas of inevitable human and organic progress, including Christian-based beliefs in the divine perfection of species through special creations – culminating, of course, in man and the angels. As a passionate supporter of scientific naturalism, Huxley felt bound to lash evolutionary thinking as bad science.[41]

For much the same reason, he heaped praise on Charles Lyell's claims that the geological record showed no evidence of a progressive fossil evolution, one that developed in complexity over time.[42] All theories of development were, in Huxley's mind, the product of metaphysical rather than naturalistic thinking. In his own way, then – if for entirely opposite reasons – the brilliant young biologist could prove to be as much of an obstacle to the acceptance of Darwin's theory as Owen himself.

Darwin had another very good reason for needing to sway Huxley and Hooker during that Downe weekend. As well as being the two most influential scientists of their generation in their respective fields, they had begun to accumulate political clout within the wider institutional world of science. Huxley and Hooker had effectively become joint leaders of a group of young scientists who wanted to reform Britain's old guard of clerical dilettantism and entitlement.

This group had begun to form in the early 1850s as an informal circle whose members all had at least one key point in common. They saw themselves as outsiders struggling to find scientific work in the face of social and religious disadvantage. One nucleus had been the fraternity of the sea – youngish naturalists who shared the experience of having worked as surgeon's mates in the navy.

Hooker and Huxley, who became instant comrades themselves, had also befriended George Busk, a recently retired surgeon from HMS *Dreadnought* at Greenwich, who shared their exasperation with the Admiralty's parsimony. As a founder and former president of the London Microscopical Society, Busk teamed up with Huxley to pore over specimens of sea ferns collected on the *Rattlesnake*. Both also helped Jock MacGillivray publish *Narrative of the Voyage of the Rattlesnake* in 1852. Huxley provided illustrations, Busk specialist notes on Bryozoa (small, mosslike marine animals that grow in colonies).

Busk had a seaman's diffidence and honesty, and Huxley, in particular, was drawn equally to his dark-eyed, freethinking wife, Ellen. Of humble social origins herself, this clever, acerbic woman was someone the lonely biologist could confide in about his painfully deferred relationship with Nettie Heathorn.[43]

In 1851, Huxley and Hooker had chanced on another likeminded outsider at a meeting of the British Association for the Advancement of Science in Ipswich. John Tyndall, an Irish shoemaker's son, was around their age and cast in the same intransigent mould. Though reared a Protestant Orangeman, he had gradually lost his faith while following an occupational path remarkably similar to Alfred Wallace's. Having educated himself at Mechanics' Institutes, he'd worked as a surveyor, mapmaker, engineer and schoolteacher, before throwing in his job in 1848 to try to make a living by his prodigious natural talents as a physicist. Like Huxley, he admired Germany, where he'd been able to obtain cheap scientific instruction in 1850, and the two also shared a passion for the writings of Thomas Carlyle, who worshipped work and hated shams.

John Tyndall introduced Huxley and Hooker to friends who echoed his humble origins – men with Mechanics' Institute backgrounds and freethinking outlooks. Soon their circle had expanded to include a formidable group of young scientists. There was the chemist Edward Frankland, illegitimate by birth and raised by a cabinet-maker stepfather; ex-surveyor and wool-stapler's son, Thomas Hirst, who excelled at mathematics and physics; a self-taught medical biologist and builder's son, Edwin Lankester; and Aberdeen doctor-turned-botanist, Arthur Henfrey.

Had Alfred Wallace not still been sailing among the islands of the Malay Archipelago, he too might have joined this cadre of 'hard hands'. Most were conscious also of their religious handicaps. Being Nonconformists or freethinkers, they had no hope of working at institutions like Oxford or Cambridge, where Church of England religious tests remained a condition of employment. Religious resentment attracted two other science-minded recruits, Unitarian physiologist William Benjamin Carpenter, and freethinking social philosopher Herbert Spencer.

Lack of patronage had proved another stumbling block for most of their number when applying for scientific jobs. Hooker, unable to live off his father's diminished salary at Kew Gardens, had failed to secure a permanent position in 1845 at Edinburgh University. He'd toyed with the idea of applying for the curatorship of Sydney's botanical gardens, or even trying to live by journalism, but eventually embarked on another government collecting expedition instead.

Between 1851 and 1853, Tyndall and Huxley both applied for successive positions at the new colonial universities of Toronto and Sydney, but were defeated even there by the tentacles of nepotism. When Jock MacGillivray's father died, Huxley applied unsuccessfully to replace him at Aberdeen University. In mounting despair, because he knew it meant the end of his scientific dreams, Huxley considered migrating to Australia to become a squatter or to work as a brewer with Nettie's father. 'Science in England does everything but pay,' he railed to his sister Lizzie. 'You may earn praise but not pudding.'[44] He'd won the coveted Gold Medal of the Royal Society in 1852, but where had it got him?

By the time of Darwin's Downe weekend in April 1856, however, most of the Huxley-Hooker circle had managed to scramble into some sort of paid scientific work in London, where the clerical and 'old boy' monopoly was open to challenge. In 1855, Hooker had at last persuaded the government to make him assistant director to his father at Kew Gardens. Around the same time, Huxley succeeded his old mentor, Edward Forbes, as Professor of Natural History at the School of Mines in London. By tacking onto this several other poorly paid teaching jobs, he'd finally reached his salary goal of seven hundred pounds, and could write to Nettie in Sydney: 'So darling pet come home as soon as you will.'[45] They married at last on 21 July 1855, at All Saints' Church, Finchley Road.

John Tyndall, a regular dining companion at the Huxley house in St John's Wood, had also made good. During 1853–54, he acquired several modestly paid posts, as a professor at the Royal Institution and a lecturer at the London Institution. Even quiet George Busk had broken through: early in 1856, he succeeded Richard Owen as Hunterian Professor of Comparative Anatomy at the Royal College of Surgeons, another rather nominally paid position.[46]

These small successes intensified the circle's ambition to promote the influence of their number, and to remove the obstacles still in their way. Here Joseph Hooker proved every bit as tough an operator as Thomas Huxley. By April 1856, the two of them, along with George Busk, had gained footholds in key centres of educational power, having being appointed to set and supervise the scientific examinations for entry to and promotion within the British army and navy, the Indian army, the Apothecaries Company, and the University of London. The trio had also been elected Fellows of the Royal Society, and members of its influential dining and lobby group, the Philosophical Society. Now they were pushing to get their radical scientific friends elected. Around the same time, Busk was appointed Under-secretary of the Linnean Society, and Hooker made a Fellow.

Hooker, the best connected of the circle, seized every opportunity to advance his friends. He was disgusted, for example, that Huxley had been unable to join the Linnean Society because he couldn't afford the fees. As a prominent member, Hooker itched to turn the society from a fusty gentleman's club into a leading research institution with high-calibre publications, a library open to all genuine scholars, and a reputable journal wherein one's scientific originality could be established. He also wanted to publish a series of up-to-date natural-history textbooks and a revised set of examination curricula to match. He planned to write a textbook himself, on botany, and urged Huxley to 'occupy the field' of zoology.[47] Ambitious schemes like these could be greatly enhanced, Hooker believed, by getting Huxley and others of the circle into the Athenaeum Club, one of London's most potent social levers.

Despite all this, Hooker had by 1856 come to the view that the haphazard meetings of their circle needed to be regularised in some way. Darwin's forthcoming Downe weekend offered an excellent opportunity to talk over such a scheme with Huxley. 'I am very glad we shall meet at Darwin's,' Hooker wrote to his friend in anticipation. 'I wish we could there discuss some plan that would bring about more unity in our efforts to advance science. As I get more and more engrossed at Kew I feel I want the association of my brother naturalists, especially of such men as yourself, Busk, Henfrey, Carpenter and Darwin.' What they needed, he thought, was 'some place where we never should be disappointed of finding something worth going out for'.[48]

On the eve of the April weekend, there was a feeling of change in the air. Charles Lyell reported having a troubling conversation with Hooker, Huxley, Busk and Carpenter at the Philosophical Club of the Royal Society on the evening of 22 April 1856. After listening to the ebb and flow of discussion, he noted that 'the belief in species as permanent fixed and invariable . . . is growing fainter – no very clear creed to substitute'. Lyell had also noted that Hooker's work on New Zealand flora had drastically reduced the number of species recognised as such in that country, reflecting a belief that some form of evolution had taken place there from a few primitive forms.

Once such an idea was accepted for plants, Lyell knew, it would not be long before the origin of man was explained in the same way.[49]

No one thought to record the conversations that took place a few days later at Down House, the tenor of which we must guess from later, often oblique accounts by Darwin and Huxley.

Joseph Hooker, having arrived at the house early, was put to work at once answering a long list of botanical questions passed to him on Darwin's usual slips of paper – weeks later, Frances Hooker noted that her husband was still labouring at home to provide the answers. It's likely that Hooker also took the opportunity to tell Darwin of his plan with Huxley to form their likeminded circle into a more formal lobby group to reform British science. He and Huxley were already beginning to worry, for example, that Richard Owen's new position as Director of the British Museum would enable him to centralise London's scientific collections under his control. Kew Gardens was certain to be one of Owen's targets. Perhaps, too, Hooker raised the subject of getting Huxley elected to the Athenaeum Club, though we know that Darwin was later to reject the idea because it was too risky: he was afraid that Huxley's reputation would be hurt if Owen blackballed his notorious critic.[50] To fail in such an application was regarded as worse than not being nominated in the first place.

Thomas and Nettie Huxley arrived slightly later than the other guests, and were met at nearby Sydenham station with a carriage that took them the nine miles to the house in time for dinner. That evening, John and Ellen Lubbock also joined the party. Exactly what was said at the dinner table is unclear. We may guess that Darwin steered the conversation around to subjects like the hazy distinction he'd found between varieties and species of Cirripedia. This was a topic guaranteed to set Hooker ranting against the nitpicking way that some botanists liked to pamper their egos by naming new species.

Thomas Wollaston proved a disappointment at dinner by digging in his heels and declaring his unshaken belief in the divine permanence of species, notwithstanding that he'd found new insect species in Madeira – these he dismissed as isolated exceptions. To Darwin, this seemed irrational. 'Do you not feel that "your little exceptions" are getting pretty numerous?' he asked the entomologist. At the same time he privately cursed the man's theological intransigence.[51]

The next day saw business disguised as pleasure. The genial host led his guests on a tour of the gardens, greenhouses and dovecotes – a ploy aimed at instilling doubts about the fixity of species. It was Huxley's turn to become the chief focus when Darwin showed off his ninety types of pigeon, pointing out that this array of difference had been developed from a common ancestral source by skilful breeders who'd selected and exploited tiny chance variations among offspring. Later in the afternoon, Darwin interrogated each of them by turn in his study, slips of paper in hand.

Huxley, we know, was given a genial grilling. Why was he so resistant, Darwin asked, to the idea that the modern horse might be descended from the small, three-toed Eocene fossil? Did he really think there was no connection between South American llamas and armadillos and the similarly structured fossil beasts that Darwin had himself found? Why did Huxley not believe in the possibility of such intermediate groups? Above all, why did he hold to such sharp distinctions between varieties and species, even in the case of Darwin's Cirripedia evidence?[52]

Darwin listened politely to Huxley's evasive answers.[53] Even so, the hypersensitive young biologist cannot have missed his host's intention. After Huxley had left the study, Darwin jotted down his thoughts about some of Huxley's fallacies. Perhaps he raised these with Huxley in conversation later that evening – we don't know – but something at any rate shook Huxley's certainty. In his own account of the weekend, he admitted to being provoked by Wollaston's conservatism into supporting Darwin with 'a counterblast' against the clergyman's son. He also recollected that 'Hooker . . . was already . . . "capable de tout" in the way of advocating Evolution.'[54]

Someone later told Charles Lyell that wild things had been said by all the group on the subject of evolution. He reported to a naturalist relative that 'When Huxley, Hooker, and Wollaston were at Darwin's last week they (all four of them) ran a tilt against species – further I believe than they are prepared to go.' He wondered whether they had been so 'unorthodox' as to adopt Lamarck's crazy ideas about animals being able to inherit acquired traits.[55]

Darwin was sufficiently encouraged by his weekend of soundings to announce to Lyell three weeks later, on 14 May, that he'd decided to write a book on 'Natural Selection'. He nevertheless continued to scan his friends' letters and writings for any signs of a shift in their views. Thankfully the signs were there. On 16 May, the *Medical Times* reported a lecture by Huxley scorning the idea that species had originated through special creations. If one assumed instead that species arose by natural laws, he'd argued, 'there can be no doubt that some form or other of that hypothesis . . . called the "Theory of Progressive Development" will present by far the most satisfactory solution'. Huxley-watchers must have been taken aback. Was this the same man who'd torn the 1854 edition of *Vestiges* to shreds?

Early the following year, Huxley's regular column in the *Westminster Review* also echoed some of the April weekend's discussion. Wollaston did not cease to disagree with Darwin on evolution of species, nevertheless he had helped to clarify 'the chaotic assemblage of facts about animal and plant distribution'. Huxley went on to say, 'the whole subject of the influence of climate, habits of life and other external conditions, as well as the capacity for variation inherent in each type, requires a thorough re-investigation'.[56] It sounded as if he were clearing the way for Darwin's book.

For two years after that April weekend, Darwin worked steadily through his mass of notes on natural selection, drafting sections and chapters for his projected book. From time to time he sent out specialised chunks for comment to trusted friends, including Huxley and Hooker. Whether or not Huxley was a complete convert to evolution remained a moot point, but it was clear that the young tiger was beginning to enjoy goading opponents of the theory, especially, of course, Richard Owen.

In several widely publicised lectures of February and April 1857, Owen had thrown down a new gauntlet to would-be evolutionists by sharply demarcating man from the apes. A fossil discovery in France of a giant apelike creature, *Dryopithecus*, had prompted him to make the astonishing claim that a unique, god-given segment in the brain called the 'hippocampus minor' separated humans from primates. The human brain, he thundered, stood out from that of all other mammals to such an extent that he'd coined a new term for it – *Archencephala*, the 'ruler brain'.[57] In fact, claimed Owen, a human was anatomically as far removed from an ape as a chimp was from a platypus.

Darwin, perhaps remembering the bizarre little creature he'd shot in New South Wales, was appalled by Owen's arrogance and dishonesty. It was a ludicrous comparison, and Owen knew it. Huxley, though, was delighted by this fresh opportunity to take on his favourite foe. Early in 1858, he responded with a hard-hitting paper which he gave at the Royal Institution, titled 'The Distinctive Characters of Man'. In it he offered Owen a counter-comparison: men were anatomically no further from gorillas than gorillas were from baboons. All three could be considered part of the same zoological family.[58]

It was difficult to imagine a more daring and provocative claim. There seemed little doubt that the fearsome fighter Huxley had joined Darwin's fleet.

On 18 June 1858, when Charles Darwin had already drafted several chapters of his book, he received a packet in the mail. It was postmarked Singapore. His journal for that day recorded the terse words, 'interrupted by letter from A R Wallace'.[59]

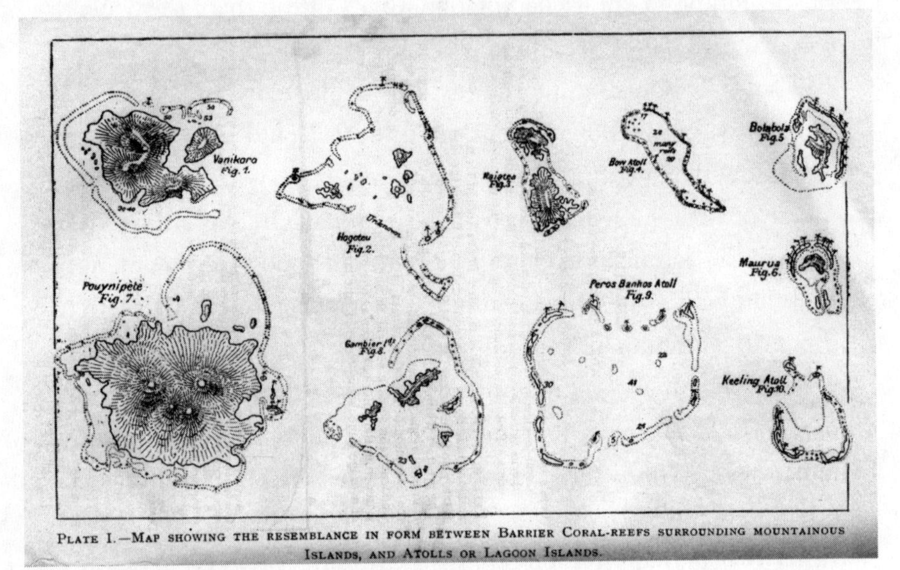


PLATE I.—MAP SHOWING THE RESEMBLANCE IN FORM BETWEEN BARRIER CORAL-REEFS SURROUNDING MOUNTAINOUS ISLANDS, AND ATOLLS OR LAGOON ISLANDS.

*Cross-section of coral reefs, with final figure showing Cocos (Keeling) Atoll, from the* Beagle's *survey*

# S.O.S.

Around nine-thirty on that morning of Friday 18 June 1858, Charles Darwin sat in his drawing room alongside Emma, reading the daily mail, as he had done with ritual regularity over the past twelve years. Tearing open a small parcel, he pulled out a covering letter from Alfred Wallace, written in Ternate, along with twenty or so pages of text on rice paper. It wouldn't have taken long to read the contents, and well before he'd finished, Darwin believed his life's work was ruined.

Later that morning, he wrote Charles Lyell a letter so terse and telegraphic that we can almost feel the effort of every word: '[Wallace] has today sent me the enclosed and & asked me to forward it to you . . . It seems to me well worth reading. Your words have come true with a vengeance that I shd. be forestalled . . . Please return me the MS which does not say he wishes me to publish; but I shall of course at once write and offer to send to any Journal. So all my originality, whatever it may amount to, will be smashed.'[1]

Much later, Darwin admitted to having been 'a little annoyed at losing all priority' when he first read Wallace's short paper, 'On the Tendency of Varieties to Depart Indefinitely from the Original Type'.[2] His friends were used to Darwin's habit of understatement, but this was an extreme example. Anger, pain and self-laceration vie with each other in all the surviving letters he wrote to Lyell and Hooker over the next month. Twenty years of backbreaking original work, squeezed out during illness and anxiety, had been eclipsed in an instant. And his labour had been dashed not by the treatise of some celebrated scholar-naturalist, but by an obscure butterfly collector in the Malay Archipelago, who'd scribbled a few pages with a minimum of facts that would now supersede the mountain of work Darwin had assembled over two decades.

Alfred Wallace was someone Darwin had never taken seriously, even in the face of warnings. He had written to Wallace in Ternate less than a year before, encouraging the enthusiastic young man to persist in his efforts as a theorist – patronising him slightly, it must be said, telling him that 'I go much further than you'. Darwin's hubris could not have brought a worse nemesis. And Wallace's little paper was good, Darwin conceded that – the man had made his case with luminous clarity and economy. Darwin told his friends that he himself could not have written a better précis of his original, unpublished and untitled draft essay of 1844 on the origin of species.

It was uncanny: the only difference between them was that he, Darwin, had focused on domestic breeding samples, Wallace on animals in the wild. In so saying, however, Darwin was understating Wallace's distinctiveness. Blinded by anguish at the thought of losing his position as the first evolutionist, all he could see was the mirror of his own mind. A deeper examination would have shown him that Wallace's theory of natural selection also differed from his own position of 1857–58 in a more substantive way. As historians Desmond and Moore pithily explain, Darwin, a liberal, believed that extinction occurred by 'a cut-throat competition among individuals'; Wallace, a socialist, believed that the environment operated to eliminate unfit groups or populations.[3]

Lyell wrote a swift reply – now lost – asking whether Darwin himself could have inadvertently given the game away. It was a tempting idea, but after thinking it over for a while, Darwin reluctantly said no. He knew his own habitual secretiveness, and anyway, Wallace's action in sending him his paper suggested innocence on that score. Lyell asked also if Darwin had anything unpublished that might establish his precedence over Wallace, something to which he could perhaps append a short new scientific sketch of his theory.

Darwin replied on 25 June that he possessed two outlines from which abstracts could be made: an 1844 sketch, the date of which could be confirmed because Hooker had read it around that time, and a more recent letter of September 1857 to the Harvard botanist Asa Gray. There was nothing Wallace had covered, Darwin claimed, that had not been dealt with more fully by the combination of these two pieces, but they were informal sketches, not really suitable for scientific publication. As for producing a new short summary of his theory:

> I should be *extremely* glad *now* to publish a sketch of general views in about a dozen pages or so; but I cannot persuade myself that I can do so honourably. Wallace says nothing about publication, and I enclose his letter. But as I had not intended to publish any sketch, can I do so honourably, because Wallace has sent me an outline of his doctrine? I would far rather burn my whole book, than that he or any other man should think that I behaved in a paltry spirit. Do you not think that his having sent me this sketch ties my hands?[4]

He concluded by asking miserably if Lyell could forward this correspondence and material to his other dear friend, Joseph Hooker.

Darwin was issuing an S.O.S. His was the letter of a decent man in an agony of indecision. He was trying to do what his gentlemanly code and instincts told him was right, but he was also fishing for a solution that would see him retain his scientific priority without losing self-respect, or that of his peers.

The following day, he sent Lyell a further letter, oscillating between these contrary feelings. What, he asked piteously, if Wallace were to respond to Darwin's publishing of a new sketch on natural selection by accusing him of taking advantage of a free and friendly communication from someone stuck in the wilds of Ternate?[5]

Anyway, writing a new sketch was now out of the question because Darwin and Emma were engulfed in a domestic crisis. Their daughter Etty, fifteen, had been showing symptoms of diphtheria for some days, and then, on 23 June, their little baby Charles contracted scarlet fever. An outbreak of the disease was sweeping the village. Poor chubby little Charles, intellectually and physically backward but full of good humour and affection, was now in serious danger. Darwin had never recovered from his daughter Annie's death four years earlier, and he worried that by marrying Emma, his first cousin, he had weakened the children through inbreeding.

Hooker, on reading Wallace's paper, wrote to Darwin proposing a joint publication and urgently requesting pieces for it. But little Charles had died the previous evening. Darwin felt 'quite prostrated, and can do nothing . . . I dare say it is all too late.' He managed to send the papers to Hooker the following day, urging him not to 'waste much time. It is miserable in me to care at all about priority.'[6] Then he washed his hands of the whole 'trumpery' business, as he and Emma prepared to bury 'our poor little darling' in the local churchyard. With a measure of bitterness, Darwin could only 'Thank God he will never suffer more in this world'.[7]

It would be a hard heart that could not sympathise with Darwin's dark night of the soul. Yet several respected historians have presented a dramatically different picture of Darwin's life during this June fortnight of 1858. They imply or claim that Darwin spent the time committing intellectual piracy and covering his tracks.[8]

Historian John Langdon Brooks, in particular, presents us with a confronting story. Darwin, he says, received Wallace's package not on 18 June, but probably a month – or at the very least, two weeks – earlier. Painstaking detective work on the Dutch East India Company's mail schedules is advanced to support the likelihood of this earlier arrival date.[9]

While Brooks concedes that Darwin's anguished first letter to Charles Lyell was not necessarily feigned, he claims that it was withheld for a considerable period. The letter, though dated the 18th, did not originally specify the month, which was added later in another hand. During the intervening period, Brooks argues, Darwin made seminal additions to his manuscript on natural selection, which were later incorporated into the published version, *On the Origin of Species*. He supposedly stooped to this heinous trick not only because Alfred Wallace threatened his place as a scientific pioneer, but also because the younger man had come up with an answer to a problem that had long eluded him.

This was the 'principle of divergence', which together with the concept of survival of the fittest explains the origin, distribution and diversity of life on the planet.[10] In essence, the divergence principle argues that varieties and species descended from a common ancestor tend to become more and more different because such diversification gives them a better chance of finding and adapting to new, unoccupied places in nature. Fitting into fresh niches enables them to increase their numbers and improve their chances of survival.

Darwin's insertion of the idea into his manuscript around this time draws heavily, Brooks argues, on Wallace's 1855 image of the process of the origins of species being like the spreading and diverging of branches on a tree, as well as on Wallace's new Ternate paper. Darwin supposedly added this reworked material some time before 18 June, and then wrote to Lyell pretending that Wallace's paper had just arrived that day.[11]

With the wave of a wand, Darwin the reclusive and invalid scholar is transformed into a pantomime villain, a Svengali figure implicated in perverted timetables, forged dates and stolen documents – all during a time of extreme grief. If the idea seems farfetched, Darwin scholars have found it necessary to expend plenty of energy and ink refuting it.

Part of the problem for anyone wanting to establish a definitive account of what happened is that too many of the pertinent letters between Darwin, Wallace and Lyell have been lost. Even so, the balance of probabilities exonerates Darwin. Given the fickleness of mail deliveries from somewhere like the Malay Archipelago, it is quite likely that Wallace's letter took an unusually long time to reach Down House. And if Darwin had really wanted to plagiarise the contents of Wallace's paper, why had he not done so and then destroyed the evidence, claiming the letter never reached him? Finally, Darwin's behaviour during this period was consistent both with his observed character and with the sequence of events relayed in surviving letters.[12]

It's difficult to establish a motive for such a risky and unethical action as Darwin is accused of. Evidence in his notebooks suggests that he began to glimpse the workings of divergence as early as 1837, when he produced a pictorial image of evolution as a branching tree. He elaborated on the idea in a variety of ways and on several occasions between 1855 and early June 1858, but the version of the principle that was eventually published alongside Wallace's paper came from a letter to the Harvard botanist Asa Gray dated 7 September 1857, well before any of the contested events. Moreover in that letter, Darwin took a distinctively ecological approach to the origin and benefits of diversity:

> The same spot will support more life if occupied by very diverse forms . . . We know that it has been experimentally shown that a plot of land will yield a greater weight if sown with several species and genera of grasses, than if sown with only two or three species. Now, every organic being, by propagating so rapidly, may be said to be striving its utmost to increase in numbers. So it will be with the offspring of any species after it has become diversified into varieties, or subspecies, or true species. And it follows, I think . . . that the varying offspring of each species will try (only a few will succeed) to seize on as many and as diverse places in the economy of nature as possible. Each new variety or species when formed, will generally take the place of, and thus exterminate its less well-fitted parent. This I believe to be the origin of the classification and affinities of organic beings at all times; for organic beings always *seem* to branch and sub-branch like the limbs of a tree from a common trunk, the flourishing and diverging twigs destroying the less vigorous – the dead and lost branches rudely representing extinct genera and families.[13]

This idea that the more diverse the organisms inhabiting a given place, the larger would be their numbers is the same version of the principle of divergence that Darwin eventually included in *Origin*. In 1857, Wallace knew nothing of it. Darwin had no need to steal from Wallace because he had already worked out his law. His place in the complete theory of natural selection was established by a letter written some ten months before Wallace sent him the Ternate paper. On the charge of plagiarism, then, Darwin has no case to answer.

And yet the miasma of conspiracy that hangs over the events of June–July 1858 is not entirely dispelled. Wallace biographer Arnold C. Brackman has levelled a second charge, one that implicates Darwin less, but his friends Lyell and Hooker centrally. Brackman accuses these two and their associates of orchestrating a plot to brush aside Wallace's contribution to the theory of natural selection in favour of Darwin's. This, Brackman says, has led to the warping of history and the eclipse of Wallace.

Thanks to the way in which Darwin and Wallace's papers were announced to the world of science, at a meeting of the Linnean Society on 1 July 1858, Darwin is today remembered as the genius discoverer of the theory, while Wallace – as novelist Margaret Drabble puts it – 'is famous for being forgotten'. Brackman argues that behind Lyell and Hooker's genteel euphemism for the joint publication, 'a delicate arrangement', lay the arrogance and manipulation of class power. Lyell and Hooker were protecting their personal friend and social equal, and defending the ranks and procedures of respectable science against a lower-class outsider in Alfred Wallace.[14]

This argument is less easy to dismiss, even though Brackman's rendering of the joint publication of Darwin and Wallace as a class conspiracy is misleading. We might come closer to understanding what happened on 1 July 1858 if we consider it in terms of a brilliant rescue, performed by friends and supporters determined to overthrow what they saw as the real class conspiracy – that of the clergy and aristocrats who were corrupting and monopolising English science.

Charles Lyell and Joseph Hooker had little time in which to rescue their friend. The scientific season was drawing to a close and most of the leading London institutions had already held their last meetings before closing for summer. Darwin himself was too sick and grief-stricken to do anything but send a bundle of informal materials unsuitable for publication. By the conventions of the day, reading out scientific papers at a meeting of a respected institution would constitute publication and so establish priority.

Even so, there were serious obstacles to their plan of a joint publication. Where to do it, for example? The Royal Society was too august, the British Association for the Advancement of Science had laid down its program months in advance, and the Zoological Society was dominated by their enemy, Richard Owen.

But luck, of a macabre kind, swung their way. The death on 10 June of Robert Brown, the noted Flinders-voyage botanist and expert on Australian flora, had led to the cancellation of a key meeting of the Linnean Society. Bylaws demanded that the society now meet hastily before the end of the season to elect another council member. This meeting was scheduled for 1 July, only twelve days after Darwin had sent his first S.O.S. to Lyell.

The Linnean Society could not have been a better venue for the publication had Hooker designed it deliberately. Actually, in an oblique way, he had done just that. He had been manipulating, reforming, and stacking the society with likeminded reformers ever since being made a Fellow in 1854. One of his aims was to give professional rigour to the society's *Proceedings* and *Journal*, so that scientists had a swift outlet for announcing discoveries. Here was his chance to prove this could be done. Lyell had little to do with the mechanics of the meeting. Unlike Hooker, he was only an occasional attendant, though his prestige as a co-sponsor of the publication of Darwin's and Wallace's papers was invaluable.

First Hooker had to force the proposed publication onto the agenda. This proved easy enough because the secretary of the zoological section of the society was his close friend, ex-naval surgeon George Busk; and the current president, Thomas Bell – investigator of the *Beagle* reptiles – regarded himself as a friend of Darwin. But it meant that the already scheduled paper-givers had to be bumped to the end of the proceedings; these included Hooker's botanical collaborator George Bentham, who reluctantly complied then later withdrew altogether.[15]

George Busk, as secretary, agreed to read out the joint papers since Darwin was burying his baby son on the day and couldn't attend. Wallace, still somewhere in the jungles of South-East Asia, was unaware of what was happening, but as he wasn't a member of the Linnean Society, he would anyway have had to ask a Fellow to read out his paper.

Next Hooker needed to ensure that this unorthodox and unscheduled intrusion into the agenda could be justified. Darwin's two papers would, of course, be read first, thereby implying his priority, but Hooker only received the materials on 30 June, the evening before the meeting. Within a few hours he had to select the best extract from Darwin's 1844 sketch, edit its rough edges, and make fine copies of all the material. He conscripted his wife Frances to transcribe Darwin's extract and to clarify some of the wording.[16] Meanwhile he drafted an introduction to the joint publication, to be signed by Lyell and himself. His words – checked by Lyell – were chosen with extreme care:

> These gentlemen (Mr Charles Darwin and Mr Alfred Wallace) having, independently and unknown to one another, conceived the same very ingenious theory to account for the appearance and perpetuation of varieties and of specific forms on our planet, may both fairly claim the merit of being original thinkers in this important line of inquiry; but neither of them having published his views, though Mr Darwin has for many years past been repeatedly urged by us to do so, and both authors having unreservedly placed their papers in our hands, we think it would best promote the interests of science that a selection of them should be laid before the Linnean Society.[17]

Both Lyell and Hooker were aware of the half-truths and verbal sleights of hand in their introduction. In fact, the two 'gentlemen's' theories were not exactly the same; original elements in Wallace's theory would now be eclipsed by Darwin having taken priority. The introduction stressed, too, that Darwin had written his material a considerable time ago, but it misled in saying that neither man had published his views on 'this important line of inquiry'. Wallace had made such a contribution in his Sarawak Law paper of 1855, which was not mentioned. Finally, the wording of the introduction implied that Wallace was a willing participant in the publication, when he knew nothing about it. As Janet Browne says, 'Wallace's work was praised then delicately moved aside.'[18]

Before presenting the joint publication, however, one final arbiter needed to be consulted. Although unable to afford the Linnean Society fees, Thomas Huxley had formed the habit prior to meetings of joining the three office-holding members of their circle – Hooker, Busk and Carpenter – in order to look over the evening's agenda and advise on tactics.[19] Whether he contributed any changes to Hooker and Lyell's introduction remains unknown, but he certainly read it and its three accompanying papers, on the morning of 1 July 1858.

The official meeting was held at Burlington House, Piccadilly, in a small room that is now occupied by lavatories. The atmosphere that evening – which saw the unveiling of one of the most important scientific theories of the modern world – could not have been duller. Only a tenth of the membership turned up – around thirty in all. The proceedings opened with Charles Lyell's respectful obituary of Robert Brown: was he conscious, one wonders, that he was symbolically burying two Australasian voyagers and sanctifying another? He and Hooker also pointed out to the Fellows that in presenting the joint publication of Darwin and Wallace they were acting not in order to establish priority for any individual, but in the interests of science as a whole. Then George Busk read out the three pieces to a stiff silence, which continued through the remainder of the program of five further papers, most of them leaden with detail.

Everyone later agreed that Darwin's and Wallace's bombshell was greeted without comment, even at the ritual tea drinking afterwards. Hooker thought that the audience had been overawed by the presence of both Lyell and himself 'as his [Lyell's] lieutenant in the affair'. They had been further disarmed by the fact that 'the subject was too novel and ominous for the old school to enter the lists before armouring'; otherwise, Hooker claimed, they would have 'flown out against the doctrine'.[20]

They might equally have been stunned by the tedium of the crammed proceedings, or simply been unable to digest the significance of what they'd heard. The zoological president, Thomas Bell, has gone down in history for uttering one of the most comically inept statements of all time when he announced in his subsequent address that 'the year [1858] has not, indeed, been marked by any of those discoveries which at once revolutionize, so to speak, the department of science on which they bear'.[21]

The theory delivered, Hooker's work was not yet over. With the help of Huxley and Busk, he quickly began work to publish the joint papers the following month, when they appeared under slightly different titles in the *Journal of the Proceedings of the Linnean Society.*

All in all, he, Lyell and friends had pulled off a superb feat. When the news reached Darwin on the Isle of Wight, where he'd fled to the beach with his family to escape from scarlet fever, he was '*much* more than satisfied'. The Linnean meeting had underscored his priority much more emphatically than he'd expected; he had imagined his abstracts would be appended to Wallace's paper. Were it not for Lyell and Hooker's 'generous kindness', Darwin said, he would have relinquished all priority to Wallace; he'd been halfway through a letter with that in mind.[22]

All the same, the deftness of the action increased Darwin's

anxiety about how Wallace might see it. Guilt and indignation jostled each other in his subsequent letter to Lyell: 'I do not think Wallace can think my conduct unfair, in allowing you and Hooker to do what you thought fair.'[23] Even so, he was anxious that Hooker write to explain the circumstances of the joint publication. 'I certainly should much like this, as it would quite exonerate me,' he pleaded.[24] Hooker immediately obliged, and Darwin included his tactful letter with one of his own to Wallace soon after. He knew, though, that he would have to wait months for a response, as the package crawled its way by steamer, prau and canoe to the far side of the world and back.

Had Darwin's friends acted immorally? Certainly they had bent the rules, to advance their friend's position at Wallace's expense, but by the conventions of their day they probably believed they had been fair to the unknown collector. Because of its author's obscurity, Wallace's Ternate paper would, unaided, have received no hearing other than in the popular journals where he ordinarily published. There, most naturalists would have ignored or dismissed it. Hooker and Lyell had given Wallace access to a major scientific institution, where he was carried on the tide of Darwin's reputation. Though Hooker and Lyell had engaged in dodgy practice, they were right in claiming that the vocation of science benefited from their intervention.

Alfred Wallace was later to make these same points himself, though he did regret that his paper had been published before it was ready.[25]

Tough-minded Joseph Hooker knew that even now the rescue remained incomplete. He was conscious that their 'arrangement' was fragile, that Wallace or others might be working on a more substantial publication, and that Darwin's cumbrous manuscript might take years to finish – if, indeed, his health allowed it. He was still recuperating from one of the most traumatic periods of his life, but Hooker pressed him to write a formal scientific abstract for publication in the Linnean Society journal.

Darwin jibbed and grumbled: hadn't he said a thousand times that an abstract was impossible without all the supporting evidence? But Hooker was remorseless. He would, if necessary, publish a special, much larger edition of the journal to carry the abstract, but it must be done.

In his heart Darwin knew his friend was right. Annoyed and weary, he dragged himself away from his therapeutic walks along the beaches and cliff-tops of the Isle of Wight to start work on the wretched abstract.

In the weeks and months following Darwin's return to Down House, Hooker began to notice that he'd uncaged a tiger. Though Darwin's stomach continued to torment him, he was writing so fast that his abstract threatened to burst the bounds of a journal, even the special issue of a hundred and fifty pages that Hooker had offered. Soon it was clear that Darwin's scholarly article had evolved into a book. He was writing what would become *On the Origin of Species*.

Hooker and others sensed that Darwin had rediscovered himself. The elephantine manuscript 'Natural Selection' had been killing him. When he'd groaned to Lyell in March 1857 that 'it is beyond my powers', he'd not exaggerated.[26] He had been so concerned about his reputation in the eyes of his family, his scientific peers and the world that he'd tried to make the work definitive – a graveyard for any scholar. Now he felt liberated. He was writing 'one long argument', without footnotes, using all the literary and visual brilliance that had gone into his *Voyage of the Beagle*.

Historian Janet Browne has given us a vivid picture of him, 'secure in his downland ship', sitting in a cabin-like office, bolt upright on his mobile chair, writing furiously on a plank laid across his knees.[27] All he needed to complete the re-enactment was a hammock and the slap and sway of the Pacific rollers. Wallace had provided the goad to action, Hooker and Lyell had forced him to show his colours, and he sensed that he had a support fleet at the ready.

If anything, Darwin's newfound vigour increased his reliance on his friends. 'You cannot imagine what a service you have done me in making me make this abstract,' he wrote excitedly to Hooker in October, 'for though I thought I had got all clear, it has clarified my brain much by making me weigh relative importance of the several elements.'[28] Every word was sent to Hooker for forensic examination, though Darwin sometimes found the scrutiny a touch excessive.[29] But his anxieties about this criticism evaporated when his friend assured him that natural selection was proving a 'jampot' for his own botanical research. 'I forgot for the moment,' Darwin wrote in relief, 'that you are the one living soul from whom I have constantly received sympathy.'[30]

Hooker meant it about the jampot. As he told Asa Gray, 'I can now use Darwin's doctrines – hitherto they'd been secrets I was bound to honour, to know, to keep, to discuss with him in private and to combat if I could, in private – but never allude to in public.' He'd had to write his own scientific work 'as if I had never heard of Natural Selection, which I have all along known and feel to be not only useful in itself as explaining many facts in variation, but as the most fatal argument against "Special Creation" and for Derivation being the rule for all species.'[31]

Under the circumstances, it was rather rich for Darwin to congratulate him that 'Natural Selection has acted as a purgative on your bowels of mutability'.[32] But it was true that the adoption of natural selection was transforming Hooker's work. While his Indian and New Zealand directories of species of 1855 had been barely touched by Darwin's ideas, his *Flora Tasmania* became a powerful case study of the operation of natural selection in Australia, written in tandem with Darwin's *Origin*.

What made Hooker's case study so compelling was the distinctiveness of Australia's botany. The country boasted the most remarkable flora on the globe. A continent with relatively uniform physical features would be expected to generate a large number of individual plants of a few major types, yet Australia turned these expectations upside down. Genera and species abounded. With this went an exceptionally high percentage of unique flora: two-fifths of Australian genera and seven-eighths of its species were found nowhere else.[33] Hooker was able to confirm these figures by drawing on data from his own voyage, on the expertise of precursors like Robert Brown, the industry of colonial collectors such as Ronald Gunn, and on Sir William Hooker's peerless herbaria at Kew.

In all, Hooker had managed to examine seven thousand out of Australia's estimated eight thousand species. He was staggered to find that they differed between the south-west and the south-east of the country, even though there were no evident geographical barriers between the two regions that would prevent the mixing of species. The southern island of Van Diemen's Land only added to the puzzle. In geological terms, it had parted from the mainland only recently, but it carried a much larger percentage of plants from both Antarctica and Europe.

In trying to explain these puzzles, Hooker conceded that he could not actually disprove the orthodox theory of special creations and permanent species. However, creationism offered him only a dead end – the unpredictable interventions of a supernatural power. He decided therefore that he would adopt a hypothesis which, though also unproven, meshed with his experience and explained his data.

Briefly, it was that nature everywhere, both in the wild and in cultivation, generated a perpetual and infinite amount of plant variation, a proliferation checked only by spontaneous cross-fertilisation and the extinction of weaker varieties through the struggle for existence. '[M]ost plants,' he wrote, 'are at warfare with one or more competitors for the area they occupy, and . . . the number of any individuals of any one species are contingent on the conditions which determine those remaining so nicely balanced that each shall at least be able to hold its own, and not succumb to the . . . smothering influences of its neighbours'.[34] Because of this perpetual struggle, weaker intermediate forms were eliminated and species took on an illusion of permanence that disguised their actual plasticity.

This continual change also influenced the geographical distribution of plants. A handful of great orders of plants like ferns, along with large groups like legumes, were dispersed almost equally over the globe. These plants tended to be simple in structure and prolific in their variations. Whenever they became isolated for any length of time, they gave rise to distinct but closely allied forms, such as the distinctive legumes of Australia. The global pervasiveness of these groups suggested that they had managed to overcome all obstacles to spreading, and were able to generate new species in competition with each other. By contrast, more specialised plants, such as the desert shrubs of Australia, were limited not by competition from other species – few of which could cope with the extreme conditions – but by factors like climate.

Darwin and Wallace's evolutionary theory, Hooker said, could help to explain the anomalies of Australian floras. This age-old continent had, over the eons, been colonised by early plant types and subsequently by migrants from Europe, whose passage could be followed in a trail of alpine species along former glacial highlands, and in the country's two most southerly corners. These corners, once separated from each other by sea, had in a relatively recent geological time been linked by the elevation of desert-like land between. In short, the south-west and the south-east of Australia carried the traces of distinct floras developed long ago in different regions of a vast but now sunken or fragmented continent.[35]

As he drafted *Origin*, Darwin was buoyed unimaginably by this support from Britain's most talented and scrupulous botanist, even though Hooker claimed that comparing his work and Darwin's was like running a ragged handkerchief up the mast beside a Royal Standard.[36] Darwin even began to worry that he had bullied his friend into intellectual compliance. Had he stifled Hooker's originality with his incessant flow of questions and suggestions?

Thinking this over, he eventually decided that Hooker could not have made his sweepingly original analysis of Australian floral origins and distributions unless he'd first abandoned the stultifying theory of permanent species. 'There is no doubt to it,' he reassured both himself and Hooker in November 1858, 'you would have arrived at the mixture [of variation and mutability] independently.'[37] As *Flora Tasmania* and *Origin of Species* marched together towards completion, Darwin felt consoled that 'they contain much the same . . . yet somehow everything is taken up from such different points of view that I do not think we will injure the originality of our respective books'.[38]

When it came to injuring books, Hooker had his own moment of horror after he stowed Darwin's *Origin* manuscript in a scrap-paper drawer and his children destroyed a large chunk of it. It seemed that the struggle for existence extended even to books. 'I feel brutified, if not brutalized,' Hooker told Huxley guiltily, 'for poor Darwin is so bad he could hardly get up the steam for what he did.' He feared that Darwin's health would give out with this extra challenge.

Huxley was reminded of what happened to the manuscript of another of his heroes, Thomas Carlyle's *French Revolution*, which had been destroyed by a maid. Thankfully Darwin possessed a copy. Huxley ordered his friend Hooker to pay penance by standing on his head in the garden for an hour.'[39]

Huxley himself did not have Hooker's long history of convergence and collaboration with Darwin, but it was precisely his record of intelligent scepticism about the ancestry and mutability of species that made his support so invaluable. Huxley was equally delighted at the way 'Wallace's impetus seems to have set Darwin going in earnest', and he began to anticipate that a fuller version of the theory would effect 'a great revolution'.[40] It is likely, too, that Huxley's bestselling *Lectures on General Natural History* of 1856 influenced Darwin's bold decision to write an abstract on natural selection that he hoped would be 'popular to a certain extent'.[41]

Darwin felt encouraged that Huxley had accepted evolution provisionally, but still doubted his friend had made a full conversion. There was no explicit evidence of evolutionary ideas in Huxley's marine work of 1858–59, which focused on the usual issues of anatomical classification. Huxley had been working on *Oceanic Hydrozoa* since his return from Australia, but lack of financial support, along with his struggle to carve out a viable scientific career, stretched the book's production over more than a decade. As he put the finishing touches to the manuscript in 1859, Huxley was simultaneously reading proof sections from both Darwin's and Hooker's new works, and he envied the freshness of their approaches.

With some trepidation Darwin sent Huxley his conclusions about embryology, which suggested that the embryo of any given species carried the imprint of its ancestral origins. His elegant description of this as 'a picture, more or less obscured, of the common parent-form of each great class of animals' brought Huxley's warm approval. Here, in the ancestry of the past, Huxley conceded, was the real source of the archetype. In his bluntest admission yet, he reported to Hooker that 'the facts seem to me to come very strong for the mutability of species'.[42]

Even so, when Huxley pointed out to Darwin that natural selection would remain open to doubt until breeders managed to produce a sterile cross, as occurred in natural species, Darwin agreed: 'It is a mere rag of a hypothesis with as many flaws and holes as sound parts. My question is whether the rag is worth anything.'[43] In fact, Huxley was itching to run Darwin's red rag up the masthead.

During the frenetic, thirteen-month gestation of *Origin*, Darwin's friendship with his two brilliant lieutenants intensified. Since the Down House weekend of April 1856, his relationship with Huxley had moved quickly to the level of personal intimacy that he'd long enjoyed with Hooker. One sign of these tightening personal bonds was Huxley's request in 1858 that Emma Darwin stand as godmother to his newest baby, Jessie Oriana. Religious sceptics though they were, Huxley and Darwin appreciated the symbolic force of connecting their families in this way. The triangular cross-flow of letters between Darwin, Hooker and Huxley also increased in tempo and warmed in tone. The three ex-sailors had become a troika.

At the same time, despite their crushing workloads, neither Huxley nor Hooker neglected their larger mission to mobilise a circle of scientific reformers in Darwin's support. John Lubbock had joined their number after the Down House weekend, and quickly proved his value by alerting Darwin to some miscalculations in his botanical arithmetic that were hampering his proofs of divergence.[44] Huxley – political to his bootstraps – negotiated a new outlet for their reformist views in the fashionable *Saturday Review*, and he, Hooker and John Tyndall were soon regular contributors. And when Huxley gained membership of the Athenaeum Club, Hooker urged his friend to pay the fees at once so that they could nominate George Busk and Tyndall to help 'swamp the parsons'.[45]

Huxley needed no encouraging. From the outset he sensed that Darwin's doctrine of transmutation had the potential, in Desmond and Moore's words, to 'split science from theology'. He could feel revolution in the air, and his anticlerical ferocity sometimes made even his friends shiver. 'If I have a wish to live thirty years,' he growled in January 1859, 'it is that I see the foot of Science on the necks of her enemies.'[46]

That same January, Darwin received a welcome letter from the Malay Archipelago. He had been fretting about Wallace's response to the 'delicate arrangement' since July the previous year. Though Wallace's letter of January 1859 is now lost, its friendliness brought Darwin a gush of relief. Far from taking umbrage, Wallace was grateful for the recognition. His decency shone, too, in a letter he sent in reply to Hooker's:

> Allow me in the first place to thank yourself and Sir Charles Lyell for your kind offices on this occasion, and to assure you of the gratification afforded me for both the course you have pursued, and the favourable opinion of my essay which you so kindly expressed. I cannot but consider myself a favoured party in this matter, because it has hitherto been too much the practice in cases of this sort to impute all merit to the first discoverer of a new fact or a new theory, and little or none to any other party who may, quite independently have arrived at the same result a few years or a few hours later . . .[47]

Because Darwin's work was far more advanced than his own, Wallace went on, 'it would have caused me much pain and regret had [his] excess of generosity led him to make public my paper unaccompanied by his own'.[48] Wallace sent more openly excited letters to his mother and Henry Bates, delighting in the double tribute of being praised by Hooker and Lyell and co-published with Darwin.[49]

Towards the end of January 1859, an invigorated Darwin in turn thanked Wallace for his generous response:

> Permit me to say how heartily I admire the spirit in which they [your letters] are written. Though I had absolutely nothing what-

> ever to do in leading Lyell and Hooker to what they thought a fair course of action, yet I naturally could not but feel anxious to hear what your impression would be. I owe indirectly much to you and them; for I almost think that Lyell would have proved right and I would never have completed my larger work; for I have found my abstract hard enough with my poor health.[50]

This praise spurred the exhausted Wallace to keep working in the Malay Archipelago for a further three years.

On 6 April 1859, almost a year after the fateful Linnean Society meeting, Darwin's follow-up letter to Wallace showed that their relationship had moved to a new phase. Tossing aside the conventions of Victorian formality, the two reserved men had quickly become friends. Darwin's final paragraph contained a brief, gossipy report on how two other key scientists had responded to their joint theory: 'Hooker, who is our best British botanist, and perhaps the best in the world, is a *full* convert, and is now going immediately to publish his full confession of faith . . . Huxley is changed and believes in mutation of species; whether a *convert* to us, I do not quite know.' Then, with an uncharacteristic burst of optimism, Darwin assured Wallace: 'We shall live to see all the younger men converts.'[51]

The admiral was welcoming a new captain into his armada.

# Battle

Early in April 1860, Charles Darwin, Joseph Hooker and Thomas Huxley pored together over an anonymous review of *On the Origin of Species* that had just appeared in the *Edinburgh Review*, Victorian Britain's most influential literary periodical. It was a nice coincidence that Hooker and Huxley happened to be staying for the weekend at Down House, because the piece also included jibes at Hooker's recent essay 'On the Flora of Australia', as well as some vicious sideswipes at Huxley. It was altogether nasty. The three of them knew it must have come from the pen of Professor Richard Owen, not least because it was so full of self-praise. The man's vanity was legendary.

Darwin had been expecting and dreading this moment. Because Owen was the most famous anatomist in the country, he had naturally been sent a copy of the book, along with one of Darwin's usual self-deprecating letters. A snide reply had goaded Darwin into dragging himself to London in December 1859 for a 'long interview' with the man. Though smarmy as usual, Owen had barely contained his fury. 'He was quite savage and crimson at my having put his name with the defenders of immutability,' Darwin reported to Lyell.

Since Owen believed in a type of continuous creationism, he was outraged at having been included among the reactionary old guard of non-evolutionary scientists, even though that's exactly what he was. Darwin's hubristic ambitions, he sneered, would certainly cause the book to fail with any credible scientist, though it would no doubt be acclaimed by Darwin's 'disciples'. He'd called the latter 'your Huxleys', Darwin reported, 'with a degree of arrogance I never saw approached'.[1]

The review of *Origin* was cast in the same toxic mould. It began by citing some inconsequential passages, which Owen patronised as the book's 'real gems'. Then came the demolition. Darwin, it seemed, had set out to solve the 'supreme problem' of species' origins, which the world's greatest naturalists had long faced 'with due reverence'. Instead he'd produced a muddled farrago, not unlike *Vestiges of the Natural History of Creation* fourteen years earlier. *Origin* was the product of someone whose mind was 'not weighted or troubled with more than a discursive and superficial knowledge of nature'.[2]

And since Darwin had failed to convince readers with his facts and powers of observation, the review jeered, 'we were then left to confide in the superior grasp of mind, strength of intellect, clearness and precision of thought and expression, which raise one man so far above his contemporaries, as to enable him to discern in the common stock of facts, of coincidences, correlations in Natural History, deeper and truer conclusions than his fellow labourers had been able to reach'.[3]

No matter, the book would soon be forgotten, along with the works of 'disciples' like Hooker, who was deemed 'as short sighted as his master'. Wallace too was damned as Darwin's partner in crime, and an unnamed 'Professor' [Huxley] was ridiculed for making breathtaking anatomical blunders to ignorant popular auditors.

Finally Owen spelled out the confronting implication that Darwin had been so careful to avoid – that 'a species, such as a gorilla [be] considered as a step in the transmutative production of man'.[4]

Darwin ought to have been devastated by this assassination. The review was a distillation of the worst abuse he'd been imagining for twenty years. In the months leading up to the publication of *Origin*, he'd been so sick with anxiety that his family and friends had wondered if he would survive the ordeal. Now everything he'd dreaded had come to pass. 'It is extremely malignant, clever & I fear will be very damaging,' he told Lyell.[5]

But to his surprise, Darwin was troubled by Owen's bile for only one night. 'I have quite got over it today.' Far from wringing his hands, he was filled with a kind of invigorating holy rage. Though he resisted the urgings of Huxley and Hooker to write a stinging reply, he decided to have nothing further to do with the 'base dog' Owen; the man was a conceited liar and cheat, eaten up with spite and jealousy.

Charles Darwin, it seems, was rediscovering a toughness that his sickly and retiring persona had forgotten. The keystone of this new defiance was his faith in his friends. It was they, he told Lyell, who helped him feel 'bold as a Lion'.[6] He was even buoyed by the hope that one of Owen's accusations might be true: 'perhaps the majority, of our younger naturalists have been seduced into the acceptance of . . . "natural selection"'. Along with such young Turks, Darwin knew he could also count on at least one powerful elder. As he'd told Huxley on the eve of publication, 'I . . . fixed in my mind three judges on whose decision I determined mentally to abide. The judges were Lyell, Hooker and yourself.'[7]

Each judge had delivered his verdict and Darwin was delighted with the result. On 23 November, he told Lyell, 'I rejoice profoundly that you intend admitting [the] doctrine of modification in your new Edition [of the *Principles*] . . . I honour you most sincerely: – to have maintained, in the position of a master, one side of the question for 30 years & then deliberately give it up, is a fact to which I much doubt whether the records of science offer a parallel.'[8]

In reality, Darwin was overestimating Lyell's conversion, but he was right to assume his old mentor's support. Charles Lyell could not face the implications of natural selection for humanity's divine origins and spirit, but he never hesitated to defend Darwin's theory against detractors.

Joseph Hooker's support for the theory seemed so certain by the autumn of 1859 that Darwin almost took it for granted. Still, he was elated when his discerning friend, usually so sparing with praise, called *Origin* 'a glorious book' and said: 'I would rather be author of your book than of any other on Natural History Science'.[9] Busy as he was with his own book, Hooker had dashed off a series of letters to persuade several recalcitrant botanists, at whom he'd also directed a warm review in the *Gardener's Chronicle*.

Thomas Huxley's verdict was critical, partly because he'd been the most resistant to the theory, but mostly because of his brilliance as a communicator and fighter. Darwin thought him 'the best talker I have known', with a mind 'as sharp as a razor'.[10] Huxley, Darwin said, wrote with 'vitriol rather than ink', and could command popular and specialist audiences with equal ease. He, more than anyone, had predicted that *Origin* would start an intellectual war, and he had jumped at the chance to lead it, writing to Darwin in November 1859:

> I trust you will not allow yourself to be in any way disgusted or annoyed by the considerable abuse and misrepresentation which unless I greatly mistake is in store for you. – Depend upon it you have earned the lasting gratitude of all thoughtful men – And as to the curs which will bark and yelp – you must recollect that some of your friends at any rate are endowed with an amount

> of combativeness which . . . may stand you in good stead – I am sharpening up my beak and claws in readiness.[11]

He gave Darwin a further lift by writing appreciative early reviews in *The Times*, *Macmillan's Magazine* and *Westminster Review*. By harnessing such powerful organs of Victorian opinion, he showed that the Darwinists were a force to be reckoned with. And it was Huxley, more than anyone, who rolled out the language and metaphors of battle – broadsides, squadrons, flotillas, crusades, knights, warriors, Whitworth guns, capital hits, fusillades, slaughter, annihilation and armadas. His words were emblematic of the militant, mid-Victorian British empire in which they lived, quick to rattle sabres and dispatch battleships. And his ambitions for Darwinism and science were no less imperial in scope.

With such encouragement, Darwin grew more belligerent. Richard Owen's review stirred something in him that had been slumbering since his *Beagle* days. After all, he had once carried a musket in a seaborne invasion and ridden on horseback alongside wild gauchos. He began to speak of 'our side', to congratulate his friends on being 'a good and compact body', and to see former mentors like Adam Sedgwick as 'creationist old fogies' who could be ignored. Before long he was encouraging rather than restraining Huxley's extremism.[12] Owen's review of *Origin* was to Charles Darwin an official declaration of war.

The most celebrated battle of the evolution war took place at a meeting of the British Association for the Advancement of Science in Oxford in late June 1860. The outlines of this story became well known. At a meeting of perhaps a thousand people – comprising scientists, clergymen, politicians, members of the public and undergraduates – the Bishop of Oxford, Samuel Wilberforce, primed by his friend Richard Owen, made a light and witty attack on Darwinian evolution. With deft timing, he turned in the middle of his speech to ask Thomas Huxley, in the fruitiest of Oxford tones, whether it was on his grandmother's or grandfather's side that he was related to an ape.

Thanking God under his breath for delivering him such an opening, Huxley rose to his feet to lance the bishop. He would rather, he said, 'have a miserable ape for a grandfather' than a man 'possessed of great means of influence & yet who employs . . . that influence for the mere purpose of introducing ridicule into a grave scientific discussion'. Public scuttlebutt and popular newspaper accounts soon transformed the reply into the knockout phrase, 'I would rather be an ape than a bishop'.[13]

Historian James Moore points out that the only battle in the nineteenth century more famous than this scrap between 'Soapy Sam' Wilberforce and Thomas Huxley was Waterloo.[14] Perhaps as a result, it has become fashionable for scholars to diminish its importance by pointing out the myths in the popular account. Many argue that the incident seems trivial when compared to other, sustained and complex clashes between pro- and anti-Darwin advocates during the 1860s. Most witnesses also presented accounts at odds with the standard version of the story. And each of the leading protagonists believed they'd won the debate. Samuel Wilberforce wrote that he had thrashed Huxley, to the delight of the crowd. On the other side, a number of pro-Darwinists believed that Joseph Hooker's subsequent speech had made the greatest impact on the scientists present. Hooker himself, though not a boaster, shared this view. He gave a post-mortem of the meeting to Darwin, who'd been too sick to attend.

Hooker had almost skipped the event himself, he told Darwin, because he'd been chatting with his and Huxley's old seafaring companion, Captain Dayman, and he sensed that hours of waffle lay ahead in the meeting. Grumpily he'd filed into the crowded room, where his temper worsened at having to listen to an hour and a half of 'flatulent stuff' from 'A Yankee Donkey called Draper on "Civilization According to the Darwinian Hypothesis"'. Then along came Samuel Wilberforce.

> Well Sam Oxon [Wilberforce] got up and spouted for half an hour with inimitable spirit, ugliness, emptiness and unfairness. I saw he was coached up by Owen and knew nothing . . . he ridiculed you badly and Huxley savagely. Huxley answered admirably and turned the tables, but he could not throw his voice over so large an assembly, nor command the audience; and he did not allude to Sam's weak points nor put the matter in a form that carried the audience. The battle waxed hot. Lady Brewster fainted, the excitement increased as others spoke; my blood boiled, I felt myself a dastard; now I saw my advantage; I swore to myself that I would smite the Amelekite Sam, hip and thigh if my heart jumped out of my mouth, and I handed my name up to the President (Henslow) as ready to throw down the gauntlet . . . [S]o there I was cocked up with Sam at my elbow, and there and then I smashed him amid rounds of applause. I hit him in the wind at the first shot in ten words taken from his ugly mouth; and then proceeded to demonstrate in as few more: (1) that he could never have read your book, and (2) that he was absolutely ignorant of the rudiments of Bot[anical] Science. I . . . wound up with a very few observations on the relative positions of the old and new hypotheses and with some words of caution to the audience. Sam was shut up and had not a word to say in reply, and the meeting was dissolved forthwith, leaving you master of the field after four hours of battle.[15]

Hooker was later puzzled as to why his version of the events, supported by several major newspapers, was overwhelmed by the story of a lone, titanic struggle between Wilberforce and Huxley. The answer, of course, was that the exchange between these two, small and particular though it was, ballooned into myth, thereby supercharging its effect. Myths generally work to tell a society what it most wants to know; they also symbolise the ideals and aims of spiritual or temporal hierarchies. To achieve this, they need to be fluid and baggy enough to carry multiple meanings to ranges of people. It helps, too, if they fit familiar storylines.

The Oxford debate had all these ingredients. It was like a chivalric romance, with the bishop as villain and Huxley the young knight, or vice versa if you were a creationist. The whole episode was shot through with symbols, which is why the cartoonists had such a field day.

First, there was the University of Oxford, bastion of conservatism and religion. Then the bishop, well fed and worldly, looking exactly like a stock anticlerical caricature – the type of 'three bottle parson' who'd long been the delight of popular radical and dissenter cartoonists. His opponent, lean, spare and dark, could have been a Puritan prophet, or perhaps a devil incarnate. Ironically, though, it was the bishop who resorted to levity and salacious innuendo, while Huxley, the secular moralist, trumpeted science's sacred commitment to the truth. Multiple issues could be seen in their confrontation. Wilberforce was the hunt-dog of creation scientists like Owen, Huxley the bulldog of the absent Darwin. It was Tory churchman versus freethinker, age versus youth, creation science versus evolution, Church versus science, religion versus reason, spirit versus matter – the list of symbolic battles catered for most segments of the reading public.

The incident convinced Darwin that his hopes of gaining support for the theory of natural selection must come from 'the young men growing up and replacing the older workers'.[16] War, symbolic and actual, was now declared. As the conservative *Athenaeum* newspaper put it, the bishop's and Darwin's parties 'have each found foemen worthy of their steel, and made their charges and counter charges very much to their own satisfaction and [to] the delight of their respective friends'.[17]

At the time, it looked like a pretty uneven contest. The debate had thrown a handful of junior scientists into combat with the established Church, one of Victorian Britain's grandest social and political institutions, an instrument of the state, the backbone of the constitution, and a repository of immense financial, social and intellectual power. Darwin was right, though, to stress the advantage of his circle being 'a good and compact body'. The exceptional cohesion of his small group of followers came in part from their shared wish to defend an illuminating scientific hypothesis.

But there were also other connective tissues between them, related only indirectly to Darwin's theory. Hooker needed help to defend Kew Gardens from the predatory ambitions of Owen, and he wanted Darwinist naturalism to improve the scientific status of botany, which was still looked down on as the preserve of amateur enthusiasts. Huxley wanted the introduction of a scholarly and ethical form of science, open to merit; he wanted the overthrow of the clergy, aristocrats, and social climbers like Owen who stood in his way; and he wanted to use Darwinism as a 'Whitworth gun', to bring about a 'New Reformation' that would sweep away 'the scum of rotten hypocritical conventionalism which clogs art, literature, science and politics'.[18]

Theirs was an awesome responsibility, Huxley reminded Hooker in a rallying letter of December 1860. 'You and I, if we last ten years longer . . . will be the representatives of our respective lines in the country. In that capacity we shall have certain duties to perform, to ourselves, to the outside world, and to Science.' Two men built 'on the high pressure tubular principle like ourselves' had no choice but to fight or burst.[19]

The bonds of friendship between the men were indeed like those of war veterans. Sailing and adventuring in wooden ships on the far side of the world was one adhesive, another was a kinship of personal loss. Darwin had passed through hell when his little Annie died in 1850, and Hooker had done everything in his power to comfort him. A decade later, Darwin admitted to Huxley, he still could not think of Annie 'without tears rising in my eyes', though the pain had become 'tenderer' with the years.[20]

On 13 September 1860, Huxley also found himself holding a dead child in his arms. Little Noel, aged six, had succumbed to scarlet fever. Nettie was numb and broken with grief; Huxley roared out his anguish: 'we have lost our little son, our pet and our hope'. He felt the four corners of his house to be 'smitten', and 'I could fancy a devil scoffing me'. Agonised letters to his friends helped see him through the misery, but Nettie's depression could not be alleviated, even by the birth of another child, to whom Hooker and Darwin both stood as godfathers. It was Emma Darwin who began Nettie's recovery, by inviting her and the children to visit their quiet, invalid-filled house. Here the two wives discovered a solace in reading Tennyson out loud to each other.[21]

Then, in October 1863, it was Hooker's turn to begin a black-tinged letter, with the words 'Dear Old Darwin, – I have just buried my darling little girl and read your kind note . . . It will be long before I cease to hear her voice in my ears, or feel her little hand stealing into mine; by the fireside and in the Garden, wherever I go she is there.'[22] A month later, his grief at the death of little Minnie remained completely raw: 'I shall never cease to wish my child back in my arms as long as I live.'[23] Men joined by such suffering could not easily be pulled apart.

In the spring of 1862, the circle welcomed a new fighter. Alfred Wallace's figure when he arrived back in England attested to his strength of will. Skeletal, his face emaciated by recurrent malaria, and his body covered in boils, he returned to London after eight years in the Malay tropics. He had undertaken more than fourteen thousand miles of rough sailing and collected more than twelve thousand specimens. It was some months before he could find the strength to visit Darwin, who impatiently awaited him at Down House. In the summer, however, Wallace eventually felt well enough to pay the call, and he received a warm welcome from the whole Darwin family.

Wallace had trumpeted his views on *Origin* to everyone he met. He'd told his boyhood friend George Silk that 'Mr Darwin has given the world a new science, and his name should in my opinion, stand over that of every philosopher of ancient and modern times. The force of admiration can go no further.'[24]

The admiration was mutual. Over the next decade, Darwin and Wallace forged a relationship so intimate it has been compared to a love affair.[25] Without any false deference, Wallace always saw himself as Darwin's subaltern. 'As to Charles Darwin, I know exactly our relative positions, and my great inferiority to him,' he told the Christian socialist Charles Kingsley in May 1869. 'I compare myself to a Guerrilla chief, very well for a skirmish or for a flank movement, and even able to sketch out a plan of campaign, but reckless of communications and careless about Commissariat: while Darwin is the great General, who can manoeuvre the largest army, and by attending to his base operations and forgetting no detail of discipline, arms or supplies, leads on his forces to victory.'[26]

On his first meeting with Darwin, Alfred Wallace had looked more like Rip Van Winkle than a guerrilla leader. Though only thirty-nine, he sported a long white beard and prematurely greying hair. The scientific scene had changed unimaginably in his absence, and he needed to be brought up to date with the realities of the evolution war. He was incredulous when Darwin told him that Owen had written the anonymous piece oozing self-praise in the *Edinburgh*

*Review*, and had also orchestrated the Bishop of Oxford's additional attack on *Origin* in the *Quarterly Review.*

It wasn't long, however, before Wallace was exposing the zoological errors of one of Darwin's persistent attackers, Reverend Samuel Houghton of Dublin.[27] This he followed with such a steady flow of pro-Darwin reviews, papers and articles that Hooker and others dubbed him 'Darwin's good knight'.

Charles Darwin's London friends welcomed Wallace with equal warmth. Lyell thought his 1858 paper on natural selection better than Darwin's, and Huxley was busting to enrol him as an ally. In his bestselling book of 1862, *Man's Place in Nature*, Huxley drew heavily on Wallace's published work on the human-like structure and behaviour of the orang-utan.[28] Wallace was further flattered that Huxley admired his biogeography so much that he coined the term 'Wallace's Line' to describe the Asia–Australia faunal divide. Though Wallace was too shy to enjoy the starchy atmosphere of dining with the Lyells, he felt 'quite at ease' in the domestic rough and tumble of the Huxley household, 'which is what happens to me with few persons'.[29] Young Leonard Huxley romped around with him, and Australian Nettie displayed none of the intimidating Victorian social formalities that he found so bewildering.

Huxley briefed Wallace, too, on how ancestral relations between apes and mankind had become the central battleground of the anti-Darwinists. Owen and the conservative press harped on the issue incessantly, knowing it was the point most likely to produce moral revulsion. Huxley explained how Owen, having had his divine 'hippocampus' in the human brain exposed as a sham, was now trundling out sensational stories about gorilla sexuality and violence from a dodgy French-African adventurer named Paul du Chaillu. Huxley had devoted so much time to showing up Owen's anatomical errors that he'd terminated their most recent exchange with the words 'Life is too short to occupy oneself with the slaying of the slain more than once.'[30]

Even so, Wallace watched in awe at a meeting in Cambridge in 1863 of the British Association for the Advancement of Science, in which Huxley systematically demolished Owen's assertions about ape anatomy. Charles Kingsley echoed the opinion of most attendants when he commented that Owen had deserved 'the thrashing he got'. But it was *Evidence as to Man's Place in Nature* – Huxley's little 'monkey book', as Darwin called it – which completed Owen's evisceration. Short, flinty and vastly popular, it was written for working men and it bluntly called Owen a liar. It also contained a pictorial demonstration of the anatomical similarities between primates and humans that drove home Huxley's argument more vividly than a hundred learned papers.[31]

Huxley obviously needed no help from Wallace 'to nail [Owen] out like a kite to a barn door', but the same did not apply to a formidable new foe who emerged in 1863 to attack Darwinism from a different quarter. In that year, a Dr James Hunt started the Anthropological Society, which claimed to have founded a new form of science. Medically trained, this brash, flamboyant young man aimed to professionalise the study of man and race, which was then the province of a missionary-dominated organisation called the Ethnological Society. Hunt proposed using science to prove that humans had originated from multiple primate ancestors and then developed into the existing white, red, black and yellow races. Among these, Hunt and his cronies believed, whites were the most superior, blacks the most primitive, and the two races should never mix.

The speedy growth of the Anthropological Society reflected both Hunt's charisma and a political climate conducive to theories of white domination. The 1850s and 1860s saw a spread and hardening of British imperial attitudes in places like India, South Africa, the West Indies and Australia. The United States was split by the

Civil War between the north and south over the issue of black slavery, which attracted corresponding adherents in Britain. Supporters of slavery dominated the Anthropological Society. In the name of science, Hunt and his bohemian mates like the explorer Richard Burton also promoted a culture of masculine racial eroticism in a kindred organisation called the Cannibal Club, complete with skulls, shrunken heads and African masks.

Most of the 'Anthropologicals' disliked Darwinian theory because of its assumption that races were not separate biological species, but closely related human varieties evolved from a single ancestral source. While even liberal Darwinists like Huxley and Wallace were not immune from the prejudices of their day, they saw civilisation as an attainable social stage for all races, irrespective of their biological endowments. Such an idea was anathema to Hunt.

Against this background, in 1865, Huxley read a controversial paper, 'Emancipation – Black and White', in which he attacked both black slavery and the subordination of women. He cited a natural law 'that no human being can arbitrarily dominate over another without grievous damage to his own nature'.[32] Huxley loathed Hunt so much that he stopped attending meetings, but his co-campaigner Wallace chose to carry on. He endured ridicule for papers that praised the abilities and achievements of Malays and Negroes and attacked the record of British settlers in the empire. Citing examples of frontier violence against Australian Aborigines, Wallace declared, 'The white men in our Colonies are too frequently the real savages, and require to be taught and Christianized quite as much as the natives.'[33]

His most important theoretical contribution, however, came with the paper, 'The Origin of Human Races and the Antiquity of Man, Deduced from the Theory of Natural Selection', which he read to the Anthropological Society in 1864. He argued that 'a grand revolution in nature' had taken place in the remote past, at a time when the human mind evolved consciousness. Until then, man's body and brain had been shaped by the struggle for survival that nature visited on all species, including man's primate ancestors. But the advent of 'mind' changed all this.

First, it drove man to diversify into a variety of races spread across the globe. Second, it deflected the forces of evolution away from the human body, which could now combat environmental pressures with tools, clothes, housing, fire and crops. From this point the human body remained essentially static. Natural selection's transformative energies shifted to the sphere of morals, where progressive improvements could continue indefinitely. 'By his superior sympathetic and moral feelings [man] becomes fitted for the social state; he ceases to plunder the weak and helpless of his tribe; he shares the game which he has caught with less active or fortunate hunters, or exchanges it for weapons which even the weak or deformed can fashion; he saves the sick and wounded from death.' Under this progressive drive, Wallace proclaimed, man would eventually become a single unified species once more, and life 'a bright paradise'.[34]

Hunt scorned the paper as a wet product typical of one of 'Darwin's disciples', but Wallace was proud to be so called, and they to have him. This was the first serious work of social Darwinism to break the link between biological struggle and human ethics, and its importance is often overlooked. Three decades later, the idea would resurface in Huxley's most famous lecture, 'Evolution and Ethics'.[35]

Hooker certainly noticed: he told Darwin that he was 'amazed . . . at [Wallace's paper's] excellence. It never struck me to account for the fixity of man as Wallace has done.'[36]

It was a work of 'remarkable genius,' Darwin agreed. Wallace himself remained diffident. 'My great fault,' he confessed to Darwin, 'is haste. An idea strikes me, I think it over for a few days, and then write away with such illustrations as occur to me while going on. I therefore look at the subject almost solely from one point of view.'[37]

Hasty it might have been, but it inspired Darwin to begin his own exploration of the issue of man and evolution, which appeared six years later as *The Descent of Man*. Wallace, it seems, had again been the catalyst for Charles Darwin's genius.

In 1864, the battle for natural selection entered a decisive new phase through the founding of the X Club. On 3 November, Huxley and Hooker summoned six friends to dinner at Saint George's Hotel, Albemarle Street, near Piccadilly Circus, a short distance from the Royal Institution. Their purpose was to found a dining club, which would meet at six or six-thirty on the first Thursday of each month. They would then stroll around the corner to the meeting of the Royal Society, where all but one were members.

It was a familiar eight: Huxley, Hooker, John Tyndall, George Busk, George Frankland, Herbert Spencer (the non-member of the Royal Society), Thomas Hirst and John Lubbock. The following year, they were joined by Tyndall's friend William Spottiswoode, a distinguished mathematician and printer, making nine in all.

A dining club was not exactly a new idea for them. Most had been dining and caucusing informally for some years. Wallace was probably too shy and reclusive to be invited, and Lyell was too representative of the older generation. Darwin's illness precluded him, but he followed their doings through Hooker and Huxley. 'Beyond personal friendship,' Thomas Hirst later wrote, 'the bond that united us was devotion to science.'[38] He meant science as a profession open to merit; science as a naturalist practice free from theological entanglement; science as a means of social, moral and national regeneration; science as a synecdoche for Darwinism.

Together, Huxley claimed, the men covered enough of the key scientific fields to make an encyclopaedia. Hooker, as we have seen, had earlier thought about forming such an organisation, but nothing came of it. Since then, however, a war over the origin of species had broken out.

The immediate incentive for founding the X Club, whose name was suggested by Ellen Busk, came from Hooker's and Huxley's recent difficulties in trying to get Darwin awarded the Copley Medal, the Royal Society's highest honour. Even though Huxley, Lubbock and Busk, along with Benjamin Carpenter, were on the society's council, they'd been foiled in their first attempt the previous year by the machinations of Owen, who nominated the old creationist warhorse Adam Sedgwick instead. This year, under the subtler leadership of Hooker and Busk, they'd succeeded; Darwin was awarded the medal in November 1864, though the citation made no mention of *Origin*. Several of the old buffers on the council asked what this fellow Darwin had written.

Huxley was livid at the book's exclusion, and bulldog-like, he hounded the minutes secretary until it was included in the published proceedings. But the experience had shaken him: informal action was clearly not enough. Planned cohesion was needed if the larger war was to be won. And so the X Club was born.

Its rapid success came from being a flexible, multipurpose, tightly knit body. Historian Roy Macleod has called it a 'social network of science' and 'an open conspiracy'. Huxley himself thought of it as 'a scientific caucus or ring'.[39] The members were a freemasonry, a lobby group, a friendly society, an 'invisible club', 'a tribe', an 'intellectual aristocracy', and more, but the cement that bound them together was sociability.[40] Meetings often generated gossip, teasing and fun, as well as serious strategy. Like so many successful Victorian associations, it was strengthened by the support of 'Yvs' (wives) who organised convivial outings, readings of Tennyson, moonlit trips on the Thames, picnics at Leith Hill in Oxford, and drives to the beeches at Dropmore, followed by boating on the river.[41]

The X Clubbers dined at each other's houses, went on hikes and holidays together to Wales or Derbyshire, climbed strenuously in the Swiss Alps, gathered as families, and shared the rites of passage of their children. They corrected and discussed each other's work and offered support in times of difficulty or financial need. When Huxley was so exhausted from overwork in 1873 that his doctors predicted a complete physical and emotional collapse, it was his X Club friends and Darwin who raised the money for him to take a three-month holiday in Egypt.

The X'ers made friendship a machine of war, and harnessed its energy to storm and infiltrate the fortresses of science. They were a meritocratic 'conspiracy', each member boasting real talents, and credentials won in the face of social disadvantage. Collectively they were unstoppable. They nominated each other for awards, refereed each other for jobs, published each other's work, sponsored each other's lecture tours, awarded each other grants, and circulated each other's achievements. Their honours list tells the story. Over the next two decades, five X Clubbers received the Royal Medal of the Royal Society, three the Copley Medal, one the Rumford Medal. Six were presidents of the British Association for the Advancement of Science, three were associates of the Institute of France, and three – Hooker, Spottiswoode and Huxley – served as successive presidents of the Royal Society, between 1873 and 1885. With the exception of Spencer, they all rotated through other influential positions in the society – secretary, foreign secretary and treasurer. Over the same period, the number of clergy elected to the society fell from eight to five percent.[42]

The X'ers also founded or dominated major scientific journals. Huxley made the *Reader* 'an organ' of the Darwinists from 1863, and in November 1869 co-founded *Nature*, which became one of the most prestigious research journals in the world. Seven members of the X Club published at least one article in its opening issues of 1869–70.[43]

Five years after the difficulties over Darwin's award of the Copley Medal, the X'ers were easily able to secure a Royal Medal for maverick Alfred Wallace. By this time they were operating as a well-oiled team. Hooker raised the issue of Wallace's nomination, Huxley and Hooker lobbied members for support, Darwin wrote a reference praising his 'extraordinary talents', and Huxley eventually made the formal nomination. On 30 November 1868, Wallace was awarded this equivalent of a scientific knighthood in Burlington House, next door to where he and Darwin had co-published a decade earlier.[44]

The X'ers also realised that in order to substantiate Darwinian evolution and turn Victorian Britain into a scientific society, they had to go beyond the institutions of science. They set out to gain access to every major establishment of national power – government, parliament, universities, schools, the Admiralty, the arts, the Church.

The last, in particular, was critical. One of their shrewdest, and often underappreciated, achievements was to ally themselves with liberal Anglican clergymen who were engaged in their own war to purge militant Tories like Samuel Wilberforce from the offices of the Church. These allies included Canon Farrar at Harrow; George Rolleson at Christchurch; Arthur Stanley – the younger brother of the captain of the *Rattlesnake* – in the Philosophical Society; and the liberal Biblicist from Natal, John William Colenso, all of whom wined, dined and plotted with the X'ers to promote science in schools, appoint liberal clerics to universities, and resist the inquisitorial attacks of Church Tories.

Thomas Huxley even contrived a respectable philosophical position for sceptically minded X'ers. It enabled him, as Lenin later quipped, 'to hide his materialism under a fig leaf'. Elected to the illustrious Metaphysical Society in 1869, where he mixed with religious dignitaries of every stamp, Huxley coined the term 'agnostic', for philosophical neutrals who regarded the verdict about a divine creator as unproven either way. With a sigh of relief, secular-minded liberal intellectuals all over the country fitted themselves with agnostic fig leafs.

The X Club members received a boost to their efforts to lobby parliament and government in 1870 with the election of Lubbock – now Sir John – as a Liberal Member of Parliament. A colleague of the party leader, William Gladstone, Lubbock was able to exercise behind-the-scenes influence to help Hooker, Huxley and the other X'ers defend Kew Gardens from the grasping hands of Richard Owen at the British Museum and the bloody-minded bureaucracy of the government's Commissioner of Works, Acton Smee Ayrton. After a tough campaign, in which the X Club lobbied fiercely, the 'brute' Ayrton was defeated in 1872 and lost his seat in parliament. Kew's autonomy was now secure.[45]

For Joseph Hooker, Director of Kew Gardens since the retirement of his father in 1865, the victory lifted a threat that had hung over his entire tenure in the job. Now he could concentrate on making Kew the greatest hub of botanical science in the empire.

At what point could the Darwinists declare the invasion over, the battle won? Social revolutions rarely have neat concluding dates. Some would argue, and with good reason, that Darwin's armada has not won yet, especially in the United States.[46] However, within the more limited context of Victorian Britain, the key battles for acceptance of Darwin's theory and its associated scientific culture were over in a surprisingly short time. Huxley estimated in 1859 that victory would take seven years, and his guess was close.

There were signs as early as 1866 that the tide had turned. In that year, Joseph Hooker addressed an evening meeting of the British Association, using an allegory that revealed more than he realised. Adopting the persona of a David Livingstone figure, he reported having witnessed missionaries at Oxford in 1860 preaching a brave new doctrine that was reviled by the gathering. Six years later, a meeting of the same group was 'applauding the new creed'. Observers reported that Hooker's missionary analogy was greeted with laughter and thunderous cheers.[47] Perhaps the most revealing aspect of his parable, however, was that the language of battle had turned into the language of religious denominationalism. Darwinism was no longer an invasion force, it had become an established 'creed'.

Thomas Huxley's public nicknames tell a similar story. During the 1860s, young scientists called him 'the Captain', while his co-workers and students at the School of Mines dubbed him 'the General'. During the 1870s, he became 'Darwin's Bulldog', and then, more significantly, 'Pope Huxley'. By that time he'd achieved a type of papal dominion over science. Every new lecture, article and book of his became an encyclical. His bestseller of 1870 was even entitled *Lectures and Lay Sermons*. The man once excoriated as the 'Devil's Chaplain' had become a priest of science.

Indeed, as Principal of the South London Working Men's College in Blackfriars, Huxley was lecturing respectable working men near the spot where the original 'Devil's Chaplain', Reverend Robert Taylor, had forty years earlier preached mock sermons to worker radicals.[48] During the 1870s, too, as Adrian Desmond has observed, 'Darwinism' – the creed of a specific scientific party – gave way to the more universal term 'evolutionism'.[49]

Joseph Hooker was a barometer of the changed climate. In 1868, he again addressed the British Association, this time as president. Reviewing the progress of Darwinism after a decade, he concluded that few scientists now openly rejected the theory. They'd been persuaded, he said, by the likes of the 'true knight' Alfred Russel

Wallace, who'd just delivered a drubbing in the *Quarterly Review* to a leading Darwinist critic, the Duke of Argyle.[50] Following that meeting, Wallace himself complained to Darwin that he could no longer find any credible opponents.

Soon afterwards, Wallace delivered his own, rather paradoxical demonstration of how securely natural selection had taken root. As he and Darwin had pushed their theory into uncharted waters during the mid-1860s, they'd sometimes disagreed on details. Darwin admired Wallace's explanation of the bright colouration of some butterflies and birds as a deterrence mechanism that had evolved because such colours were associated with poisonous or distasteful properties.

They disagreed, however, over Darwin's secondary evolutionary mechanism of sexual selection. While Darwin believed that the struggles by males to win female mates through anatomical and behavioural displays had influenced the evolution of species decisively, Wallace thought it a much less potent force. Even so, neither Darwin nor his other disciples had seen these differences as politically damaging. Darwin, Huxley and Hooker were deeply proud of the success of Wallace's naturalist travel book of 1869, *The Malay Archipelago*, generously dedicated to Darwin. Nothing as brilliant had been written, the three believed, since *Voyage of the Beagle*. But in that same year, Wallace suddenly moved in a direction that was to damage his scientific standing fundamentally.

First he announced his conversion to the fashionable vogue of spiritualism, an action which has troubled his biographers ever since. He'd been driven in part by emotional need, seeking consolation for the unexpected break-up of his first love affair and for the death of his mother. But in becoming a spiritualist, Wallace was doing nothing out of keeping with his popular radical ideals. As a committed socialist, mesmerist and phrenologist, he was continuing a familiar intellectual trajectory. All over Britain and America, old radicals and utopians were making the same conversion to a form of demotic belief that they held to be both scientifically plausible and free from the control of establishment culture.[51]

Wallace's Darwinist friends thought his conversion to spiritualism silly, but a related step proved more serious. In 1869–70, he published a review and paper suggesting that the law of natural selection could not explain the evolution of human consciousness. He thought it more plausible that a higher power or 'Intelligence' had guided the growth of the human mind to specific ends.[52]

Using the Darwinian dictum that naturally selected change must be immediately beneficial to the species involved, Wallace argued that it could not therefore have produced the large brain of primitive man, whose prime needs were for strength, swiftness and cunning. The massive early brain of humans only began to fulfil its potential use much later, when social organisation generated the need for sophisticated abstraction, aesthetic appreciation and moral behaviour. Wallace surmised, therefore, that man must have been programmed for civilisation by some higher intelligence.[53]

Darwin was devastated by this inexplicable retreat. He compared it to child murder, and scored the text of Wallace's 1870 paper with marginal notations of 'no, no'. Darwin and his friends also feared that this *volte face* would encourage their creationist enemies, but here their worries proved groundless. Wallace's backsliding on this single aspect of natural selection damaged only his own reputation. By 1870, Darwinian theory could withstand even a perceived defection by its co-discoverer.

The most potent test of Darwin's theory came just a year later, when he published *The Descent of Man*. In it he advanced at much greater length the same argument about the origin and evolution of man that he'd put to Wallace in 1870. It was wholly unnecessary, he said, to introduce a higher intelligence to explain man's development; the incremental gradations of evolutionary complexity from primate to human over eons were enough to account for consciousness, signs of which could be seen in both his own and Wallace's investigations of animal behaviour over many years.

To Darwin's surprise, the spectre of the ape ancestor no longer generated countrywide panic. He encountered little or no odium, and earned huge acclaim for an argument that a decade earlier he'd been too afraid to make.

Charles Darwin knew now that his invasion was complete, even if one of its heroic captains had withdrawn from the final battle. The last paragraph of *The Descent of Man* carried his deepest belief:

> We must . . . acknowledge, as it seems to me, that man with all his noble qualities, with sympathy which feels for the most debased, with benevolence which extends not only to other men but to the humblest living creature, with his god-like intellect which has penetrated into the movements and constitution of the solar system – with all these exalted powers – Man still bears in his bodily frame the indelible stamp of his lowly origin.[54]

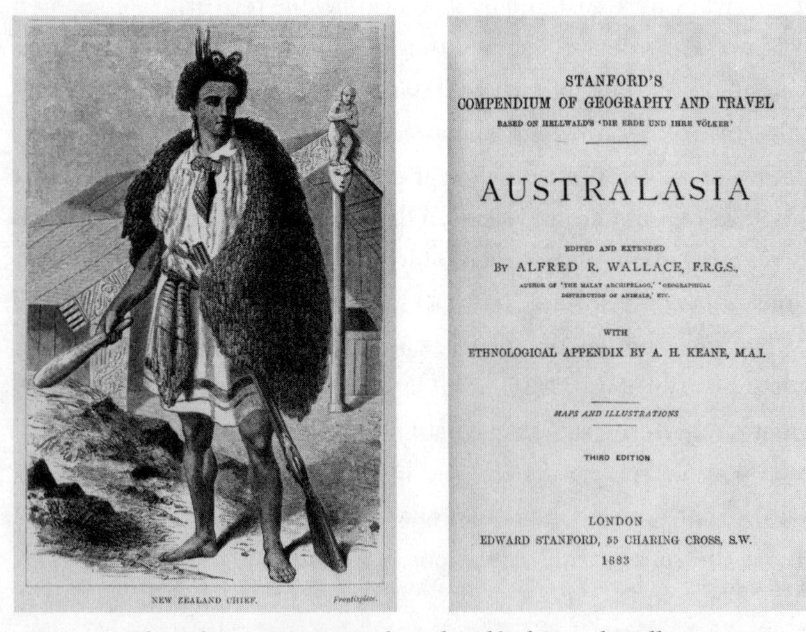


STANFORD'S
COMPENDIUM OF GEOGRAPHY AND TRAVEL
BASED ON HELLWALD'S 'DIE ERDE UND IHRE VÖLKER'

AUSTRALASIA

EDITED AND EXTENDED
BY ALFRED R. WALLACE, F.R.G.S.,
AUTHOR OF 'THE MALAY ARCHIPELAGO,' 'GEOGRAPHICAL DISTRIBUTION OF ANIMALS,' ETC.

WITH
ETHNOLOGICAL APPENDIX BY A. H. KEANE, M.A.I.

*MAPS AND ILLUSTRATIONS*

THIRD EDITION

LONDON
EDWARD STANFORD, 55 CHARING CROSS, S.W.
1883

*The title pages to* Australasia *by Alfred Russel Wallace*

# Epilogue: A Pension for a Captain

In late December 1879, Charles Darwin received a visit from Arabella Buckley, whom he'd met when she was a secretary to Sir Charles Lyell, now buried in Westminster Abbey. At the age of seventy, Darwin was more reluctant than ever to deal with these interventions from the larger world. His energies had evaporated in recent months, and his few outside interests with them. But this meeting stung him in an unexpected way.

A fellow spiritualist, Arabella Buckley had long been one of Wallace's confidantes, and she knew he was now in parlous financial circumstances. Despite the happy life Wallace had shared with his botanically minded wife, Annie Mitten, since their marriage in 1866, he was struggling to educate his two children, Violet and William, 'in the most economical way'.[1] Worried about the family's impending ruin, Arabella decided to beg Darwin for his help. Could he and Hooker – now Sir Joseph – use their influence to obtain for their friend a small government pension?

Like his father, Alfred Wallace had proved hopeless with money.

He'd squandered most of the earnings from the sales of his collections and books on investment mirages touted by over-persuasive friends. A foolish wager with a man who turned out to be mentally disturbed had sucked him into a morass of litigation that further drained his resources. Moved by a reclusive impulse similar to Darwin's, he'd shifted from house to house looking for a quiet retreat in the countryside, usually losing money each time. And his awkward blend of honesty and idealism had put off every potential employer. A succession of failed applications for minor scientific positions had forced him to look for administrative work in several schools and libraries, but always with the same result. Most recently, he'd missed out on a promising job at Epping Forest by publishing before his interview an ambitious proposal for conservationist reforms. The commissioners had been relieved to choose an innocuous landscape gardener for the position.

Wallace had kept his family from penury only by inspired hackery; he'd churned out reviews, social journalism and lectures, edited other writers' work, and scribbled books for money, including a travel guide of Australasia in 1879. But now even his fluent pen could no longer keep up with the bills. Wallace typically blamed no one but himself. He believed he was 'constitutionally lazy, without any of that fiery energy and intense power of work possessed by such men as Huxley'.[2]

Darwin was genuinely distressed at the news of his friend's decline. Wallace had rather dropped out of his consciousness in recent years and he'd had no idea of the family's plight. Wishing 'most heartily' for their wellbeing, he dashed off a letter asking Hooker to endorse the pension idea.

To Darwin's dismay, the swift and unexpected reply revealed Sir Joseph at his most starchy: 'Wallace has lost caste considerably, not only by his adhesion to Spiritualism, but by the fact of his having deliberately and against the whole voice of the committee of his section of the British Association, brought about a discussion on Spiritualism at one of its sectional meetings . . . Then there is the matter of his taking up the Lunatic's bet . . . and pocketing the money.'[3]

Wallace had accepted a bet from a flat-earth fanatic that he could not prove the world was round, unaware that a madman's logic could never be satisfied with mere scientific evidence. Wallace took the matter to court, where the battle grew so protracted that his legal fees surpassed the amount of the wager. Hooker, getting the details wrong in the long tortuous process, believed that Wallace had taken the madman's money, and thought he deserved his current penury for having dragged the reputation of science in the mud with his legal squabbles. Moreover, even if they did request a pension, Hooker added, they would be honour-bound to reveal Wallace's dalliance with spiritualism in case it brought the government into disrepute as well.[4] Hooker did not mention, though he probably also had them in mind, Wallace's other embarrassing involvements with socialist, feminist and land-nationalisation campaigns.

Darwin was saddened by this response, but knew when he was beaten. Hooker in such a frame of mind was immoveable. Darwin wrote to Arabella Buckley explaining the hopelessness of the case. She was not surprised. Wallace's maverick spirit had cost his friends any amount of anguish over the years.

But what had happened to down-to-earth Joe Hooker, the man who once praised Wallace as a 'good knight'? He was now of course Sir Joseph, a pillar of the state, and had only recently finished serving a term as President of the Royal Society. Being the successful Director of Kew Gardens had also put him in command of a vast network of economic and medicinal resources. Prime ministers, heads of state and Whitehall mandarins treated him with awe and respect.

Still, no one could accuse him of having lost his old reformist spirit. He'd raised ten thousand pounds to lower the Royal Society membership fees for poor naturalists like Alfred Wallace. He'd also remained in close and respectful contact with his old colonial helpers, like Ronald Gunn, to whom he'd dedicated his *Flora Tasmania*.

Not even Hooker's elevation to knighthood in 1877 can be blamed for his anti-Wallace outburst. Hooker had twice earlier turned down the offer of a knighthood for services to the colonies (KCMG) because of his dislike of empty honorifics, and he eventually accepted only after reassurance that it was being awarded for his botanical work in India. '[I]t is not given by favor,' he'd told Darwin half apologetically, '. . . and it has a flavor of hard work under difficulties, of obstacles overcome, and of brilliant deeds that is very attractive.' No, it was not snobbery that had prompted Hooker's snooty remarks about Wallace, but rather a passionate desire to protect the hard-won victories of Darwinian science.[5]

Wallace, meanwhile, knew nothing of this rebuff. He decided to improve his shaky finances by writing a popular book about all the discoveries made by himself and his voyager comrades and synthesising them under the subject of biogeography. Published by Macmillan in early autumn, 1880, it was called *Island Life: The Geographical Distribution of Animals.* Ironically, Wallace dedicated it to Sir Joseph Hooker.

The first half of the book contained a lucid explanation of the subject's theoretical challenges. The biogeographer needed, Wallace argued, to first map the global distribution, variation, modification and dispersal of species and groups, and then to analyse the way geological change had affected continents and islands. He had then to explore the nature, causes and effects of climate changes, and finally to determine the rate of organic development in relation to geological time.

The second half of the book examined a series of case studies involving the trickiest known 'insular floras and faunas'. One by one, Wallace re-traversed the islands that he and his three fellow voyagers had visited mid-century, including the Galapagos, St Helena, the Sandwich Islands, Borneo, Java, the Madagascar group, Celebes, New Zealand and Australia. References to the published work of Sir Joseph Hooker were studded throughout the book, particularly in the last two chapters – 'The Flora of New Zealand: Its Affinities and Probable Origin' and 'On the Arctic Element in South Temperate Floras'. Both chapters compared the way island ecologies like New Zealand's differed from continental ecologies like Australia's. The emphasis Wallace gives to the actions of major climatic change remains pertinent today.[6]

*Island Life* became the foundation stone of the modern discipline of biogeography and began the resuscitation of Wallace's scientific reputation. Without intending it, Wallace also revived the admiration of his fellow Darwinists. Darwin himself praised him unreservedly: 'I have now read your book, and it has deeply interested me. It is quite excellent, and seems to me the best book which you have ever published . . .'[7] Just as they had done in the heroic days of the 1860s, he and Wallace plunged into detailed exchanges of facts and theories.

*Island Life* also cracked a tougher nut. Even before he finished reading it, Hooker wrote excitedly to Darwin saying that Wallace had already brushed away more 'cobwebs' than any previous biogeographer. Hooker thought the work 'splendid' and could hardly believe that such a brilliant man could also be a spiritualist.[8]

Still shrewd despite his diminished energies, Darwin sensed the opportunity to revive the pension idea. This time he decided to direct operations through Huxley. Not only had he always been personally closer to Wallace than any of them, he also knew what it was like to stare poverty in the face. Despite Huxley's scientific eminence, he was struggling to support a family that included both his own large brood and the sons and daughters of his impoverished sister. The strain had taken its toll. By some strange affinity, he'd joined Darwin in suffering from what he called chronic 'nervous hypochondria' and 'dyspepsia' of the stomach. Years of compulsive overwork had battered and scoured that once raptorial face.

Throughout it all, Nettie had been his rock, seeming to grow wiser and stronger with each vicissitude. She had helped her husband, 'the human tugboat', to mellow. Huxley rarely fired broadsides these days, though his tongue had lost none of its bite. When Bishop Samuel Wilberforce – Soapy Sam – died earlier that year after falling from a horse, Huxley quipped: 'Poor dear Sammy! His end has been all too tragic for his life. For once reality & his brains came into contact & and the result was fatal.'[9]

Huxley was the best person to lead the campaign for Wallace's pension in another sense, too, because he'd become the official historian of the evolution war. In 1880, he delivered a lecture to a packed Royal Institution crowd entitled 'The Coming of Age of the Origin of Species'. With his usual flair he celebrated the maturity of Darwin's revolution and recast the history of the previous twenty-one years so as to accentuate the overthrow of recalcitrants and reactionaries like Richard Owen. The victors tell the story.[10] The whole vocation of science, Huxley suggested, had come of age in tandem with Darwin's great theory.

Huxley never doubted that Wallace deserved the thanks of his country, and he assured Darwin that he would put all his 'might, amity and authority' behind the cause.[11] Arabella Buckley drew up a memorandum of Wallace's achievements and Huxley tweaked it to suit the prejudices of Queen and Parliament. It listed Wallace's three greatest scientific accomplishments: his astounding Malay Archipelago collections, his original contribution to the discovery of the theory of natural selection, and his brilliant application of that theory to the geographical distribution of animals.

To these Darwin added a fourth – Wallace's contribution to explaining the mechanisms of colouration in animals. Neither Darwin nor Huxley saw any need to mention Wallace's non-scientific involvements in spiritualism or socialism. It was irrelevant and would be tactically foolish. The evangelically minded Prime Minister Gladstone, whom Huxley thought as slimy as Richard Owen, was known to dislike radical and freethinking scientists.

There remained prickly, scrupulous Joseph Hooker to be won over. *Island Life* had changed his attitude wonderfully, but he was still troubled by the scandal over Wallace's wager with the madman. Huxley appeased him with a white lie. He said that Wallace had given the proceeds of the bet to charity, when they'd actually vanished into the pockets of the lawyers.

Then the armada sprang into well-practised action. Huxley suggested Darwin send a personal note to Gladstone. 'Mr G. can do a thing gracefully if he be so minded.'[12] But Darwin did much more: he became a powerhouse, finding the energy from somewhere to throw himself into the kind of political lobbying he'd always loathed. He wrote letters to everyone who might exercise influence, even persuading Wallace's old creationist foe the Duke of Argyll to put his signature on the memorial beside Darwin's own.

Gladstone received the request early in January 1881 and he knew enough of Darwin's status in the world of science to act on it quickly. Meanwhile Darwin fretted; he could think of nothing else. 'I hardly ever wished for anything in my life as for its success,' he told Arabella Buckley.[13]

On 7 January 1881, he received among the daily influx of letters at Down House an official envelope from Whitehall. Her Majesty and Mr Gladstone were glad to offer Mr Wallace a Civil Service pension of two hundred pounds a year for his services to science. 'Hurrah!, hurrah!' Darwin wrote to Huxley the same day. 'Read the enclosed. Was it not extraordinarily kind of Mr Gladstone to write himself at the present time? . . . I have written to Wallace. He owes much to you. Had it not been for your advice and assistance, I should never have had courage to go on.'[14]

Alfred Wallace was touched and relieved beyond measure at this 'very joyful surprise'.[15] Darwin's letter reached him on his fifty-eighth birthday, and he said he had never received a better present. He need feel no further anxiety about giving his children a proper education. He owned a modest cottage at Godalming, not far from Downe, and his occasional income from writing would be enough to ensure a comfortable life. 'There is no one living to whose kindness in such a matter I could feel myself indebted with so much pleasure and satisfaction,' he told Darwin from the bottom of his heart.[16]

Wallace gave similar thanks to Huxley, who was equally elated at the news. He was involved in his own celebration at the time: the silver anniversary of his wedding to Nettie. All 'the chicks and their husbands' were gathered round the dinner table. Huxley later confessed to Darwin that he liked such anniversary occasions, 'being always minded to drink my cup of life to the bottom, and take my chance of the sweets and bitters'.[17]

Three months later, Charles Darwin wrote his last letter to Alfred Wallace. By this time Darwin was deep into his last book, on earthworms, and it was bringing him a quiet pleasure, not least because of the active involvement of his children. He was finishing where he had begun, observing the minute activities of creatures living in interdependence with the English countryside. Like tiny coral polyps building walls that protected coral atolls from the southern waves, worms, while struggling to survive and perpetuate themselves, were simultaneously instilling life into the earth on which every being depended.

Actually, worms were now almost Darwin's only pleasure. In other respects, he confessed to Wallace, 'life has become wearisome to me'.[18] Even scenery no longer elated him, and he could hardly find the strength to hobble along his favourite Downe walk. He was feeling ready to go.

Charles Darwin could not predict what the future would bring his three fighting captains. He did not know that Thomas Huxley's name would become synonymous with the word 'science', or that Huxley would grow increasingly troubled by the social ideas being disseminated under the name of Darwinism.

Huxley's last great lecture, 'Evolution and Ethics', was delivered in 1893 to Soapy Sam's successors at Oxford University. It made a plea for all Darwinists to remember that 'Social progress means a checking of the cosmic process at every step and the substitution for it of the ethical process, the end of which is not the survival of those who may happen to be the fittest . . . but of those which are ethically the best.'[19]

Darwin would have been less surprised, though, to learn that the grizzled old bulldog died fighting, still defending agnosticism against the theologues to the very last. On 29 June 1895, Thomas Huxley died of a heart attack; he was later buried in Marylebone cemetery alongside his infant son Noel, with three simple lines composed by Nettie on his headstone:

> Be not afraid you waiting hearts that weep
> For still he giveth his beloved sleep
> And if an endless sleep, He wills, so best.

Darwin was not to know, either, that the pension he'd obtained for Alfred Wallace would sustain his friend for thirty more years – of campaigning for social justice and writing daring books on every facet of science, even including the biology of other planets. We can imagine how satisfying Darwin would have found the scene at the Linnean Society in 1908, when on the commemoration of their joint publication Wallace was issued with a gold Darwin-Wallace medal stamped with their bearded faces. At the ceremony, Sir Joseph Hooker gave a gracious speech on being presented with a silver copy of the same medal.

By this time, Wallace's eccentricities had been forgotten not only by Hooker, but by British science as a whole. The tireless and good-humoured old collector had become a national treasure, and was awarded the Copley Medal and an Order of Merit in the same year. At the golden jubilee of the publication of *Origin* in 1909, Wallace gave the official Royal Institution lecture. Though his voice was feeble, his passion was undimmed. Darwin, he said, had given the world a theory which surpassed all others in its understanding of our planet:

> its persistence in ever-changing but unchecked development throughout the geological ages, the exact adaptations of every species to its actual environment both inorganic and organic, and the exquisite forms of beauty and harmony in flower and fruit, in mammal and bird, in mollusk and in the infinitude of the insect-tribes; all of which had been brought into existence through the unknown but supremely marvellous powers of Life, in strict relation to that great law of Usefulness, which constitutes the fundamental principle of Darwinism.[20]

There was something apt, too, about the fact that these two tough southern voyagers, Alfred Wallace and Joseph Hooker, both died peacefully in their sleep after long lives: Hooker in December 1911, Wallace two years later. Hooker was buried at Kew Gardens and Wallace at the local cemetery in Broadstone, the last town he'd lived in. Both had resisted efforts to book them a place in Westminster Abbey, though they could not escape the trappings of fame altogether. On 1 November 1915, amid the carnage of a different and bloodier war, they were commemorated with a plaque in the Abbey.

But all this lay ahead, and Charles Darwin would not live to see it. In December 1881, he experienced his first seizure. A succession of smaller heart attacks followed, one of the most severe in March 1882 when he was taking his daily stroll along the Sandwalk, where for forty years he'd watched ants engage in their everyday struggles and triumphs. On 19 April 1882, Darwin died.

The next time his three fighting captains met, it was to take the handles of his coffin in Westminster Abbey, as their admiral set off to join his beloved earthworms.

# Notes

*Abbreviations:*

**AJJH** Joseph Dalton Hooker, Antarctic Journal, 18 May 1839–28 March 1843, Typescript copy, JDH/1/1, Archives, Royal Botanic Gardens, Kew

**ARW** James Marchant, *Alfred Russel Wallace: Letters and Reminiscences*, New York, Arno, 1975

**AUTOBIOGRAPHY** Nora Barlow, *The Autobiography of Charles Darwin, 1809–1882*, New York and London, Norton, 1958

**BEAGLE** Charles Darwin, *Voyage of the Beagle*, London, Penguin, 1989

**CCD** Frederick Burkhardt & Sydney Smith, *Correspondence of Charles Darwin*, 9 vols., 1985–1994

**CDLL** Francis Darwin, *Life and Letters of Charles Darwin*, 3 vols., London, Murray, 1887

**DARWIN'S DIARY** Randall Keynes, *Charles Darwin's Beagle Diary*, Cambridge, Cambridge University Press, 1988

**DHNS DAR** Darwin, Huxley and the Natural Sciences, Unit Five: Scientific Papers and Correspondence of Charles Darwin, c. 1830–1882, from The Darwin Papers at the University Library, Cambridge, Thomson Gale microform, mfm 99. DAR 215–216

**HCE** Thomas Huxley, *Collected Essays*, 9 vols., New York, Appleby, 1896–97

**HH CORR** Correspondence of Thomas Huxley and Henrietta Heathorn, 1847–54. In the Thomas Henry Huxley Papers, Imperial College of Science and Technology; and in the microform collection, Darwin, Huxley and the Natural Sciences, Unit 2, mfm 054

**HUXLEY'S DIARY** Julian Huxley, ed., *T. H. Huxley's Diary of the Voyage of the Rattlesnake*, New York, Doubleday, 1936

**JMA** Wallace, Journals and Notebooks from the Malay Archipelago, (Archives of the Linnean Society of London)

**KEW** Kew Gardens Archives

**LJT** A. S. Eve & C. H. Creasy, *Life and Work of John Tyndall*, London, Macmillan, 1945

**LLCL** Charles Lyell, Life, *Letters and Journals of Sir Charles Lyell*, 2 vols., London, Murray, 1881

**LLJH** Leonard Huxley, ed., *Life and Letters of Sir Joseph Dalton Hooker*, 2 vols., London, Murray, 1918

**LLTH** Leonard Huxley, ed., *Life and Letters of Thomas Henry Huxley*, 2 vols., London, Macmillan, 1900

**MA** Alfred Wallace, *The Malay Archipelago*, Periplus reprint, 2000

**ML** Alfred Wallace, *My Life: A Record of Events and Opinions*, 2 vols., London, Bell, 1905

**NRA** Henry Bates, *The Naturalist on the River Amazons: The Search for Evolution*, Narrative Press reprint, Santa Barbara, 2002

**NTTA** Alfred Wallace, *A Narrative of Travels: On the Amazon and Rio Negro; With an Account of the Native Tribes, And Observations on the Climate, Geology and Natural History*, London & New York, Ward, Lock & Co., 1889

**ODNB** *Oxford Dictionary of National Biography*, http://www.oxforddnb.com

**PRINCIPLES** Charles Lyell, *Principles of Geology*, London, Penguin, 1997

## Prologue

1 'Funeral of Mr Darwin, Westminster Abbey, April 26th, 1882, Order of Procession', DHNS, DAR 215, p. 23.This includes a drawing of the coffin pallbearers' positions. George Darwin reminded Huxley about the rightness of having Wallace as a pallbearer. DHNS, DAR 215, v. 1, pp. 49, 50–53.
2 'Funeral of Mr Darwin', DHNS, DAR 215, v. 2, p. 11.
3 *The Times*, 21 April 1882, DHNS, DAR 215.
4 'Darwin's Home', DHNS, DAR 215, v. 2, p. 11.
5 Adrian Desmond & James Moore, *Darwin*, London, Penguin, 1992, p. 664.
6 DHNS, DAR 215, vol. 2, p. 11.
7 Desmond & Moore, *Darwin*, p. 670.
8 Ibid., p. 664.
9 DHNS, DAR 215, vol. 1, pp. 213–16.
10 DHNS, DAR 215, vol. 1, pp. 43–45; 25 April 1882, pp. 46–48.
11 J. Hooker to William & George Darwin, 29 April 1882, DHNS, DAR 215, vol. 1, pp. 225–28.
12 'Death of Mr Darwin', DHNS, DAR 216, vol. 2, p. 160.
13 *The Times*, n.d., cited in Adrian Desmond's 'Introduction', DHNS
14 Desmond & Moore, *Darwin*, p. 670.
15 *British Medical Journal*, 29 April 1882, DHNS, DAR 216, vol. 2, p. 106.
16 'The Death of Mr. Darwin', *The Nation*, 4 May 1882, DHNS, DAR 215, vol. 2, p. 92.
17 Donald Worster, *Nature's Economy: A History of Ecological Ideas*, Cambridge, Cambridge University Press, 1994, p. 147.
18 Ibid., pp. 182–84.
19 *Graphic*, 1 July 1882, DHNS, DAR 215, vol. 2, p. 77.
20 Rev. Farrar to George Darwin, 28 April 1882, DHNS, DAR 215.
21 Huxley to Francis Darwin, 21 April 1882, DHNS, DAR 215, vol. 2, pp. 21–24.
22 Huxley, 'The Funeral of Mr. Darwin', *Nature*, quoted in *The Times*, DHNS, DAR 215, vol. 2, p. 3.
23 Ross A. Slotten, *The Heretic in Darwin's Court: The Life of Alfred Russel Wallace*, New York, Columbia University Press, 2004, p. 267.
24 Hooker, 'Charles Darwin', *Nature* 26 (22 June 1882), p. 171, DHNS, DAR 215, vol. 2, p. 66.
25 J. Hooker to William & George Darwin, 29 April 1882, DHNS, DAR 215, pp. 225–28.
26 LLJH, vol. I, p. 161.
27 'Mr. Darwin's Work', *St James Gazette*, 21 April 1882, DHNS, DAR 215, vol. 1, pp. 2, 30.

## The Prodigal Son

1 Autobiography, pp. 24, 39.
2 Ibid., p. 27.
3 Ibid., pp. 28–37.
4 Ibid., pp. 41–42.
5 Ibid., p. 45.
6 Ibid., p. 47.
7 Ibid., p. 40; Gilbert White, *The Natural History of Selborne* (1789), Oxford, Oxford University Press, 1993; Richard Mabey, *Gilbert White*, London, Profile, 2006.

8 Autobiography, p. 39.
9 Ibid., pp. 53–54.
10 S. M. Walters & E. A. Stow, *Darwin's Mentor, John Stevens Henslow, 1796–1861*, Cambridge, Cambridge University Press, 2001, pp. 78–107.
11 Nora Barlow, ed, *Darwin and Henslow: The Growth of an Idea: Letters 1831–1860*, London, John Murray, 1967, pp. 34–37.
12 John & Mary Gribbin, *Fitz-Roy: The Remarkable Story of Darwin's Captain and the Invention of the Weather Forecast*, London, Hodder, 2003, pp. 41–45.
13 Gribbin, *Fitz-Roy*, pp. 46–48, 51–56.
14 Fergus Fleming, *Barrow's Boys*, London, Granta, 2001, pp.1–12; Christopher Lloyd, *Mr Barrow of the Admiralty: A Life of Sir John Barrow*, London, Collins, 1970, pp. 89–111.
15 Gribbin, *Fitz-Roy*, pp. 80–82.
16 Gribbin, *Fitz-Roy*, pp. 48, 56, 82–86.
17 Barlow, *Darwin and Henslow*, pp. 29–30.
18 DHNS, Hooker, 'Charles Darwin', *Nature*, 18 May 1882, p. 51.
19 Autobiography, p. 50.
20 Ibid., p. 41.
21 Ibid., p. 23; Desmond & Moore, *Darwin*, pp. 34–40.
22 Autobiography, pp. 59–60.
23 Walters & Stow, *Darwin's Mentor*, p. 7.
24 Jason Williams, 'Introduction', Alexander von Humboldt, *Personal Narrative of a Journey to the Equinoctial Regions of the New Continent*, London, Penguin, 1995, p. lix.
25 Humboldt, *Personal Narrative*, pp. 22–44.
26 Janet Browne, *Voyaging*, pp. 133–35.
27 CCD, vol. 1, pp. 121–22.
28 Ibid., pp. 125.
29 Desmond & Moore, *Darwin*, pp. 90–91.
30 Walters & Stow, *Darwin and Henslow*, p. 36.
31 Autobiography, p. 61.
32 CCD, vol.1, p. 142.
33 Ibid., pp. 177–78.
34 Ibid., p. 176.
35 Ibid., p. 150.
36 Ibid., p. 155.
37 Darwin's diary, p. 13.
38 Ibid., p. 11.
39 CCD, vol.1, p. 187.
40 Darwin's diary, pp. 16–17.
41 Autobiography, p. 66.
42 Darwin's diary, p. 18.

## The Philosopher at Sea

1 Darwin's diary, p. 8.
2 Ibid., p. 35.
3 Ibid., p. 16.
4 Browne, *Voyaging*, p. 178.
5 Patrick O'Brian, *Master and Commander* series, New York, Norton, 1970.
6 Darwin's diary, p. 41.
7 Ibid., p. 13.

8 He'd even set himself the task of learning French and Spanish, improving his maths and 'a little classics', plus reading the Greek testament on Sundays. Darwin's diary, p. 13.
9 Worster, *Nature's Economy*, p. 132.
10 Darwin's diary, p. 67.
11 Peter Gautrey, 'Introduction', DHNS, p. 6.
12 Darwin's diary, pp. 19–20.
13 Ibid., p. 23.
14 DHNS, 'Charles Darwin', *Nature*, 15 June 1882, p. 26.
15 Darwin's diary, p. 41.
16 Ibid., p. 48; Humboldt, *Personal Narrative*, p. 43.
17 Darwin's diary, p. 40.
18 Ibid., pp. 17–18.
19 Nicholson, 'Historical Introduction', *Personal Narrative*, London, Penguin, 1995, pp. ix–xxiii; James Paradis, 'Darwin and Landscape', *Historical Annals of the New York Academy of Sciences*, 360 (1981), esp. pp. 92–97.
20 Worster, *Natural Economy*, pp. 132–37.
21 Darwin to Henslow, 18 May 1832, CCD, vol. 1, p. 238.
22 Autobiography, pp. 65–66.
23 CCD, vol. 1, p. 202.
24 Jocelyn Hackforth Jones, *Augustus Earle: Travel Artist*, London, Scolar Press, 1980; Susanna de Vries-Evans, *Conrad Martens: On the Beagle and in Australia*, Brisbane, Pandanus, 1993.
25 Gribbin, *Fitz-Roy*, p. 128.
26 Howard E. Gruber & Valmai Gruber, 'The Eye of Reason: Darwin's Development During the *Beagle* Voyage', *Isis*, 53 (1962), p. 188.
27 Darwin's diary, pp. 6, 80.
28 Ibid., p. 8.
29 CCD, vol. 1, p. 221, fn. 2.
30 Ibid., pp. 218–21.
31 Darwin's diary, p. 62.
32 CCD, vol. 1, p. 250.
33 Darwin's diary, p. 91.
34 Ibid., p. 97.
35 Ibid., p. 81.
36 Ibid., p. 130.
37 Ibid., p. 132.
38 Ibid., p. 445.
39 Browne, *Voyaging*, p. 220.
40 *Beagle*, pp. 74–76, 84–85.
41 Ibid., pp. 88-91; Browne, *Voyaging*, pp. 255–63.
42 *Beagle*, pp. 96–97.
43 Ibid., p. 97.
44 Ibid., p. 163.
45 Browne, *Voyaging*, p. 225.
46 CCD, vol. 1, p. 303.
47 *Beagle*, p. 177.
48 CCD, vol. 1, pp. 378–80
49 *Beagle*, pp. 177–78.
50 Ibid., p. 179.
51 *Principles*
52 James A. Secord, 'Introduction', *Principles*, pp. ix–xliii.
53 Darwin's diary, p. 445.

54 *Principles*, pp. 84–102.
55 Browne, *Voyaging*, p. 189.
56 Darwin's diary, pp. 26–27; Browne, *Voyaging*, p. 186.
57 *Beagle*, p. 211.
58 Ibid., p. 232.
59 Ibid., pp. 237–39.
60 Ibid., p. 249.
61 CCD, vol. 1, p. 445.
62 *Beagle*, pp. 253–54.
63 Browne, *Voyaging*, p. 293.
64 Ibid., p. 296.

## Islands on His Mind

1 Fanny Owen to Darwin, 16 September 1831, CCD, vol. 1, p. 169.
2 Darwin's diary, pp. 380, 442.
3 Ibid., pp. 381, 388, 395.
4 Ibid., pp. 410, 413, 426.
5 D. R. Stoddart, 'Darwin, Lyell and the Geological Significance of Coral Reefs', *British Journal for the History of Science*, 9 (1976), pp. 203–34.
6 Larson, *Evolution's Workshop*, pp. 41–58.
7 Gruber & Gruber, 'The Eye of Reason', p. 191.
8 Darwin's diary, pp. 353–54.
9 Ibid., p. 356.
10 Frank J. Sulloway, 'Darwin's Conversion: The *Beagle* Voyage and its Aftermath', *Journal of the History of Biology*, no. 15 (Fall, 1982), pp. 337–51; Stephen Jay Gould, *The Flamingo's Smile: Reflections in Natural History*, London, Penguin, 1985, pp. 347–49.
11 Larson, *Evolution's Workshop*, p. 74.
12 Darwin's diary, p. 365.
13 CCD, vol. 1, pp. 452, 464.
14 Ibid., pp. 471–72.
15 Darwin's diary, p. 376.
16 Ibid., p. 366.
17 Ibid., p. 380.
18 Ibid., p. 385.
19 Ibid., pp. 383–84.
20 Ibid., p. 95.
21 Ibid., p. 403.
22 Ibid., pp. 409–10, 413.
23 Ibid., p. 398.
24 Ibid., p. 399.
25 Ibid., p. 408.
26 Patrick Armstrong, *Charles Darwin in Western Australia: A Young Scientist's Perception of an Environment*, Perth, University of Western Australia Press, 1985, p. 18.
27 Darwin's diary, p. 412.
28 Ibid., p. 398.
29 Ann Moyal, *Platypus: The Extraordinary Story of how a Curious Creature Baffled the World*, Sydney, Allen & Unwin, 2002.
30 Darwin's diary, pp. 402–03.
31 F. W. & I. M. Nicholas, *Charles Darwin in Australia*, Cambridge, Cambridge University Press, 2002, pp. 53–56.

32 Donald Worster, *Nature's Economy*, pp. 138–45.
33 Patrick Armstrong, *Under the Blue Vault of Heaven: A Study of Charles Darwin's Sojourn on Cocos (Keeling) Islands*, Perth, University of Western Australia Press, 1991, p. 85.
34 Armstrong, *Under the Blue Vault*, pp. 87–88.
35 Darwin's diary, p. 413.
36 Armstrong, *Under the Blue Vault*, p. 29.
37 Darwin's diary, p. 418.
38 Ibid.
39 Darwin's diary, pp. 417–18; Armstrong, *Under the Blue Vault*, pp. 88–105.
40 Gruber & Gruber, 'The Eye of Reason', p. 199.
41 Stoddart, 'This Coral Episode', p. 22.
42 Gruber & Gruber, 'The Eye of Reason', pp. 198–99.
43 Stoddart, 'This Coral Episode', p. 38.
44 Larson, *Evolution's Workshop*, p. 75.
45 Darwin's diary, p. 422.
46 Ibid., p. 447.

## The Puppet of Natural Selection

1 Granville Allen Mawer, *South by Northwest: The Magnetic Crusade and the Conquest of Antarctica*, Adelaide, Wakefield Press, 2006, pp. 15–16.
2 Ernest S. Dodge, *The Polar Rosses: John and James Clark Ross and Their Explorations*, London, Faber, 1973, p. 179.
3 LLJH, vol. 1, p. 41.
4 LLJH, vol. 1, pp. 41–42.
5 Mawer, *South by Northwest*, p. 16.
6 Jim Endersby, *Imperial Nature: Joseph Hooker and the Practices of Victorian Science*, Chicago & London, University of Chicago, 2008, pp. 35–36.
7 Jules Sébastien César Dumont D'Urville, *Two Voyages to the South Seas*, tr. Helen Rosenman, Melbourne, Melbourne University Press, 1992; Susan Hunt, Martin Terry & Nicholas Thomas, *Lure of the Southern Seas: The Voyages of Dumont D'Urville, 1826–1840*, Sydney, Historic Houses Trust of New South Wales, 2003, pp. 45–52; Anne Hoffman Cleaver & E. Jeffrey Stann, eds, *Voyage to the Southern Ocean: The Letters of William Reynolds from the US Exploring Expedition, 1838–42*, Annapolis, Naval Institute Press, 1998.
8 A. Hemingway, *The Norwich School of Painters*, London, Phaidon, 1979.
9 LLJH, vol. 1, p. 22.
10 Ibid., p. 29.
11 Ibid., pp. 3, 5–6; Ray Desmond, *Sir Joseph Dalton Hooker, Traveller and Plant Collector*, London, Royal Botanical Gardens, Kew, 1999, p. 12.
12 Endersby, *Imperial Nature*, pp. 128, 214–17.
13 LLJH, vol. 1, pp. 24, 27.
14 Ibid., pp. 13–14, 23, 29–35.
15 Ibid., p. 15.
16 Endersby, *Imperial Nature*, p. 33.
17 LLJH, vol. 1, p. 38; Desmond, *Hooker*, p. 14.
18 LLJH, vol. 1, p. 140.
19 Ibid., p. 42.
20 Ibid., p. 43.
21 Lecture delivered by J. D. Hooker, Royal Institution of South Wales, 17 June 1846, MS JDH/1/7, Kew Archives, fo. 353–35.
22 Ibid., fo. 353.